OPTIONS FOR DEVELOPING COUNTRIES IN MINING DEVELOPMENT

Also by Raj Kumar

THE FOREST RESOURCES OF MALAYSIA:
Their Economics and Development

FOREIGN DEBT MANAGEMENT ISSUES FOR DEVELOPING
COUNTRIES (*with Nihal Kappagoda*)

OPTIONS FOR DEVELOPING COUNTRIES IN MINING DEVELOPMENT

Grantley W. Walrond and Raj Kumar

Palgrave Macmillan

ISBN 978-1-349-18103-2   ISBN 978-1-349-18101-8 (eBook)
DOI 10.1007/978-1-349-18101-8

© Grantely W. Walrond and Raj Kumar 1986
Softcover reprint of the hardcover 1st edition 1986

All rights reserved. For information, write:
St. Martin's Press, Inc., 175 Fifth Avenue, New York, NY 10010

Published in the United Kingdom by The Macmillan Press Ltd.
First published in the United States of America in 1986

ISBN 978-0-312-58692-8

Library of Congress Cataloging in Publication Data
Walrond, G. W.
Options for developing countries in mining
development.
Includes index.
1. Mineral industries – Developing countries.
I. Kumar, Raj, 1947– . II. Title.
HD9506.D452W35 1986   333.79'15'091724   85–11918
ISBN 978-0-312-58692-8

Contents

List of Figures	viii
List of Tables	ix
Abbreviations	xi
Exchange Rates	xii
Weights and Measures Used in the Study	xiii
Preface	xiv
Acknowledgements	xvi

1 INTRODUCTION	1
Scope and Objectives of Study	2
Basis of Country Selection	3
The Mineral Sector in the Respective Countries	4
2 ORGANISATION AND ADMINISTRATION OF SECTOR	9
Mineral Rights	9
Administration	9
The Licensing System	18
Conditions of Licence for Searching for Mineral Deposits	20
Conditions of Licence for Geologically Defining and Assessing the Economic Potential of a Mineral Deposit	22
Conditions of Licence for the Development and Extraction of the Ore Deposit	25
Other Licences	26
Mineral Rights and Surface Rights	28
Nationality of Mineral Rights Holders and Local Participation	29
Termination of Rights and Dispute Settlement	30
Summary Comment	32

3 THE FINANCIAL REGIME — 34
Policy Implications of Instruments — 35
   Objectives of Government Policy — 35
Royalty — 39
Income-related taxes — 43
Resource Rent Tax — 52
Other Fiscal Dues — 55

4 ANALYSIS OF IMPACT OF REGIMES — 56
The Model — 56
Assumptions of the Model and Definitions — 57
The Results — 63

5 MINERAL DEVELOPMENT AGREEMENTS: VARIATIONS FROM THE STATUTES — 87
Issues in Mineral Development Agreements — 87
   Form of Agreements — 87
   Control and Management — 93
   Fiscal Provisions — 95
   Non-Fiscal Issues — 104
Why have Mining Development Agreements been necessary? — 105
   Issues Inherent in the Characteristics of Mining — 106
   Issues Inherent in the Structure of the Mining Industry — 107
Negotiation of Mining Agreements — 109

6 SUMMARY COMMENT AND CONCLUSIONS — 111

Appendices — 123

Appendix 1   Financial Provisions and Charges — 124

Appendix 2   Requirements in Application for a Mining Licence — 134

Appendix 3   Selected Computer Print-outs and Tables of Results — 136

Appendix 4   Depreciation Methods — 164

Appendix 5   Numerical Example of a Cash Flow Based Rent Resource Tax — 167

Appendix 6   Explanation of Some Basic Financial Concepts — 171

Bibliography — 177

Index — 185

List of Figures

Chapter 4 Analysis of Impact of Regimes

4.1 Projects flows and splits between government and investor   64
4.2 Government revenue mix based on gold price of US$450/oz for case with 100% foreign investor equity participation   67
4.3 Government revenue mix based on gold price of US$450/oz for case with 51% government fully paid equity   71

Appendix to Chapter 4

A4.1 Depreciation methods compared   166

List of Tables

Chapter 1 Introduction
1.1 Key economic indicators for mining in selected countries   8

Chapter 2 Organisation and Administration of Sector
2.1 Summary of mining legislative provisions   10

Chapter 3 The Financial Regime
3.1 Ad valorem royalty charges on selected minerals   40
3.2 Comparative basic tax rates   44
3.3 Capital allowance structure for mining

Chapter 4 Analysis of Impact of Regimes
4.1 Results of model with 100% fully paid equity by foreign investor   65
4.2 Project surplus as a percentage of total revenue   68
4.3 Difference in percentage government share of total surplus from case where it does not participate   69
4.4 Results of model with 51% government equity and 49% foreign investor equity fully paid   70
4.5 Results of model where 31% of government equity is fully paid and 20% free   73
4.6 Change in government take under different equity financing schemes   75
4.7 Results of model with 31% government equity paid out of dividends and 20% free   76
4.8 Sensitivity tests of selected regimes and price scenarios   79
4.9 Papua New Guinea: impact on additional profits tax of immediate capital write-offs   80

Chapter 5 Mineral Development Agreements – Variations from the Statutes
5.1 Selected agreements and government participation   90
5.2 A comparison of fiscal arrangements in agreements and statutory provisions   96

Appendices

A3.1 Zambia: 31% of Government equity fully paid and 20% is free – gold price at US$450 an oz 136

A3.2 Tanzania: 31% Government equity paid out of dividends, 20% free equity – gold price at US$450 an oz 140

A3.3 Malaysia: 51% Government equity and 49% foreign investor equity, fully paid – gold price at US$450 an oz 144

A3.4 Sierra Leone: 31% Government equity fully paid, 20% free equity – gold price at US$450 an oz 148

A3.5 Canada: 100% fully paid foreign investor participation – gold price at US$450 an oz 152

A3.6 Papua New Guinea: 100% fully paid foreign investor participation with capital allowance spread over life of mine – gold price at US$450 an oz 156

A3.7 Papua New Guinea: 100% foreign investor participation but with immediate capital write-off – gold price at US$450 an oz 160

A5.1 Numerical example of calculating a rent resource tax that is deductible for income tax purposes 168

A6.1 Future and present value tables 172

Abbreviations

APT	Additional Profits Tax
CFTC	Commonwealth Fund for Technical Cooperation
DCF	Discounted Cash Flow
GDP	Gross Domestic Product
IMF	International Monetary Fund
IRR	Internal Rate of Return
LIBOR	London Interbank Offer Rate
NCF	Net Cash Flow
NPV	Net Present Value
RRT	Rent Resource Tax
TAG	Technical Assistance Group

Exchange Rates

	£1 Sterling	US Dollar ($)
Sierra Leone (Leone)	6.70	6.00
Botswana (Pula)	2.09	1.70
Tanzania (Shilling)	21.20	18.10
Zambia (Kwacha)	2.63	2.42
Malaysia (Ringgit)	2.99	2.56
Papua New Guinea (Kina)	1.19	1.01
Canada (Dollar)	1.62	1.37
Australia (Dollar)	1.68	1.43

SOURCE *Financial Times*, 22 March 1985 and 26 March 1985

Weights and Measures Used in the Study

1 square mile	= 2.59 square kilometres
1 square mile (640 acres)	= 258.99824 hectares
1 acre	= 0.40468 hectare
1 hectare (10 000 sq. metres)	= 2.47106 acres
1 tonne	= 0.98420 tons
1 ounce (any weight)	= 31.1 grammes

Preface

Books concentrating on mining in developing countries have been relatively few and most of them of recent origin. These writings may be classified as those dealing either purely with the geological and extractive aspects of mining by an individual developing country or region (country studies of the various facets of mining which may be unique to those states) and those studies which concentrate on the issues relevant to mining development. This work falls in the latter category.

During our professional work, and in our academic life, we found that there was no comprehensive book to which a policy maker, student or negotiator of mining agreements could quickly refer, to guide him or her on the practical issues facing mining development. It is primarily to such persons that our work is directed.

Instead of concentrating on deducing conclusions at a theoretical level, we have developed a reasonable information base which will help those administrators who have not had the benefit of formal training in mining or economics but who are growing to appreciate the issues through experience. Others who will also find this work useful are university students pursuing careers in business or mining, where most courses do not take an integrated view of resource development.

Computer-assisted financial analysis to assist in policy formulation and negotiation is used extensively to test the robustness of particular fiscal regimes under different cost-price assumptions and the distribution of the surplus between the government and the multinational. Furthermore, the administrative, legal and institutional features relevant to exploration and development are given meticulous treatment, for these are bread and butter for the practitioners. Policy formulation guidelines are suggested both in the context of theoretical cogency and practical feasibility.

In examining all these issues, we are aware that a legitimate preoccupation of most developing countries which have either emerged, or are in the process of emerging, from colonial rule, is the redefinition, both in a philosophical and real sense, of the concept of sovereignty. For these states there is a growing awareness that the legal instruments conferring the rights to independent choice and action are only the beginning of a long and complex process of nation-building, which make interdependence an absolute necessity for survival.

Ideally, we would like to cover as many developing countries as possible. Time and resource constraints have compelled us to be biased in our selection. We have chosen Botswana, Sierra Leone, Zambia, Tanzania, Malaysia and Papua New Guinea for particular analysis. Their statutes are compared with those of two developed states, Quebec (Canada) and Western Australia, to evaluate differences in approach to mining development.

The conclusions of the study are not specific to these countries, nor are we attempting to evaluate which has a better regime. Mining regimes are continuously being reviewed and revised and some of the countries examined might have already done this during the course of the publication of this volume. The aim of choosing the various legislative schemes was to illustrate policy issues in mining with real-life examples.

Acknowledgements

This study has been made possible through the prompting guidance and support of many. However, the authors would particularly like to thank the Commonwealth Fund for Technical Cooperation (CFTC) for providing the funds and facilities to make it possible. All the members of Technical Assistance Group (TAG) and Mr Mike Faber who have contributed ideas and support in the preparation of this document are here recognised. Formal recognition must also be made of Thomas Walde and Ross Garnaut who sharpened our appreciation of the issues discussed in the study. The participation of Grantley Walrond has been made possible by the Goverment of Guyana granting him leave of absence from his formal responsibilities, and this is hereby acknowledged. Thanks are also due to June Allen, Gifty Stevens, Barbara Semple, Catherine George and Joyce Bates for their typing efforts, and to Aletha Isaacs for her assistance in editing.

The views expressed in this volume are those of the authors and not necessarily those of the institutions with which they are associated.

GRANTLEY W. WALROND
RAJ KUMAR

1 Introduction

In a world of decreasing availability of finance generally, and of mining finance in particular, the small, less developed, nation states which pinned a large part of their developmental hopes on the exploitation of their mineral resources have been set a very difficult task of attracting and maintaining that scarce capital in a capital-intensive industry. This is because mining companies as a group have the monopoly of technology and know-how. Naturally, the act of competition among states literally creates an investor's market where states can be played off, one against the other, while more concessions are won by the investor.

One set of consequences of this competition is that highly skewed agreements will be concluded in the company's favour. This, over the long term, can mitigate against availability of future finances to the under-developed countries for two reasons. The longevity of the contract will be threatened when the bargaining power swings in favour of the state as the project matures, and this in turn increases the risk index of that state and most other countries with similar characteristics.

The above underscores the necessity for less developed states to increase their information base on what others are doing, not so much as to make good their competitive attractiveness to the investor, but rather to move to a point where a broad harmonisation of policies can be developed. The instruments used by member states to attract, monitor, and regulate foreign investment, and those used essentially to foster an internal dynamic for self-reliance need to be closely studied.

The on-going international fora such as the Non-Aligned Movement, North-South Dialogue, the Law of the Sea, the Group of 77 and a host of others emphasise the global nature of the problem of world development in general and mining development in particular. The New International Economic Order, emphasising mutual interdependence between the industrialised and developing countries for world economic growth, will continue to be the framework under which the world economic machinery will be operating at least until the end of the century, if not longer. The oil price fluctuations, the exchange rate movements and the debt problems of some countries, both developed and less developed, have confirmed how interdependent the world is. Mining development will have to be looked at in the context of this interdependence.

It is in this spirit that the analysis will focus on the comparative assessment of mineral and other relevant legislation which have been adopted by the mining sector of a cross-section of Commonwealth countries.

SCOPE AND OBJECTIVES OF STUDY

Most of the less developed countries have come out of, or are in the process of coming out of, colonial rule, and a crucial issue that is grappling their minds is to redefine what sovereignty means. This is finding expression in statutes covering not only mining legislation but also investment, land, agriculture and other subjects. Many laws are being refashioned for they were written for a different era. Not only will the study examine in depth the contents of existing statutes but also the evolution of the statutes through time. It will go further by trying to spell out the price of sovereignty, that is, the bottom line figures in financial terms of majority government ownership *vis-à-vis* complete foreign investment participation. This will enable discussion on the balance between the statutes and agreements, and the range of strategies government can adopt in mining development.

As mentioned above, one of the purposes of the study is to set up an information base focusing on a wide range of issues that will confront the administrator in mining development in particular and natural resources in general. Decision-making is a time-consuming and laborious process. Many a time the administrator of a third world country is afraid of making a decision for he or she may fear being accused of giving extraordinary concessions. At times, this fear of the multinational permeates, even if it has adopted a reasonable position in the circumstances of the country and mineral being discussed. There is usually suspicion, and this suspicion largely arises because of ignorance or partial knowledge of what others are doing in the mining industry. This study, therefore, will cater for administrators in less developed countries who may not have had the benefit for formal training in business or economics but who have grown to appreciate the issues through experience. It will also benefit the university students, as the study will focus on both financial and non-financial issues facing mining in the context of development. Most business courses in universities, even in less developed countries, tend to focus very narrowly on a unidisciplinary basis, often using case studies not very relevant for the practical needs of the country.

The study therefore seeks to isolate as great a range of instruments as possible and the underlying meaning of the use of these instruments to a less developed country. The existing literature has focused on these issues but the material is in disparate sources, sometimes not easy to come by in less developed countries. This volume tries to cover in an organised and concise way a full range of concerns relevant to today's world, covering both financial and non-financial issues at length.

Computer-assisted financial capability is being built up in most less developed countries. This study has gone to some length to illustrate the technique of computer-assisted financial analysis and to highlight its importance in policy formulation and in the negotiating process with the multinational. Project analysis, which is well written in the business literature in a theoretical and company-biased manner, is done here in a dynamic context, spotlighting the key variables necessary for government decision-making. The multinationals are backed by staff well versed in such techniques, and it is prudent that governments should not only be familiar with them but also have able and competent personnel that can make judicious use of them to speed up the decision-making process, as well as to build up confidence in the decisions.

BASIS OF COUNTRY SELECTION

The six less developed countries chosen for analysis are Sierra Leone, Tanzania, Zambia, Botswana, Papua New Guinea and Malaysia. All of them are members of the Commonwealth (that is, they have a common British colonial heritage) and have achieved independence at different times over the past 30 years. Four of them come from Africa, one from Asia and the other from the Pacific. Nevertheless, the study will also draw from the experiences of other countries when discussing specific issues. There are varying shades of political complexions in the countries selected, and to some extent these are reflected in the mining statutes and agreements. The concern of this study is not on the politics but on technical issues that will face the decision maker in any less developed country. As Malaysia has a federal system of government with mining a state activity, the study focuses on the mining statutes of Perak, a leading mining state, but these are examined in the national context.

To give greater meaning and dimension to the analysis of the less developed countries, the study chose two developed states, Quebec in Canada, and Western Australia, to show how their regimes compare with those prevailing in the third world. Mining activity in Quebec is strong, and its legislation a relatively recent one. The same can be said of Western Australia. It is possible that other states could as well have been chosen in Canada and Australia, or other developed countries used, but the speed in availability of information was an important factor in the basis of selection.

THE MINERAL SECTOR IN THE RESPECTIVE COUNTRIES

This brief survey is not intended to discuss the fortunes of the mining sector of the respective countries, but purely to summarise its relative importance to the economy. The larger the contribution of the mining sector to the national economy, the larger impact it would have on economic growth, employment, foreign exchange earnings and the general stability of the country. Table 1.1 shows that the size of the mining sector in terms of contributions to the gross domestic product (GDP), export earnings and employment varies from country to country.

Looking at **Sierra Leone**, which obtained its independence in 1961, minerals contributed about US$126 million in 1980 or 11.5 per cent of the GDP. The importance of this sector in terms of economic prosperity is determined by export earnings which comprise about 75 per cent of the total. This figure has been maintained around this level since the 1970s. The principal minerals produced in Sierra Leone are diamond, bauxite, iron ore, rutile and gold. This sector took 8.1 per cent of the total employed in 1981, but revenue from the sector only comprised 1 per cent of the government revenue. Sierra Leone had a GDP per head of US$270 in 1980.

In comparison, the mineral sector of **Tanzania** (also independent since 1961) is the smallest among the countries surveyed, with a total value of US$13 million in 1978 or 0.3 per cent of the GDP. Export earnings amounted to about 7 per cent of total earnings, mainly from diamonds. Production of other minerals such as gold and tin has declined considerably, but exploration for gold and uranium is continuing. The last available figures in 1971 indicated that 1.4 per cent of the total employed was in the mining sector, and the current figure is likely to be below 1 per cent. There are also plans for an iron and steel plant.

Although not itself a copper producer, Tanzania handles about 75 per cent of the copper produced in the Zambian copper belt, most of which is transported by the Tazara railway to the port of Dar es Salaam. Tanzania's per capita GDP in 1980 was US$260.

Zambia's mineral sector has the largest impact on its economy contributing 26 per cent of the GDP and 94.8 per cent of the export earnings, principally from copper. Zambia was the fifth largest producer of world copper in 1982. Other minerals exported include zinc, lead and cobalt. In 1982 a large phosphate discovery was made at Kalwe, 220 km east of Lusaka. In 1979, about 14 per cent of the working population was employed in the mining and quarrying sector. The sector has also had an impact on government revenue, contributing 27 per cent of the total in 1978. Zambia, which had a per capita income of US$560 in 1980, obtained its independence in 1964, three years after Tanzania and Sierra Leone.

Botswana, which became independent in 1966, two years after Zambia, when it had a per capita income of US$69, had it raised to above US$1000 in 1981, largely due to mining activity. It has a relatively large and dominant mineral sector which contributed about 32% of the GDP in 1980, and 69.5 per cent of the total value of exports, comprised mainly of diamonds (60 per cent of exports), copper-nickel from the Selibi-Pikwe mine and coal from *Morupula*. In 1982 Botswana produced over 10 per cent of the world's natural diamonds. Mining provided in the same year 14 per cent of the government revenue and 8 per cent of the total employment.

Papua New Guinea is the newest state among the developing countries surveyed, having obtained its independence in 1975. Mining contributes 15 per cent of the GDP and nearly a quarter of the export earnings, which helped to give a per capita income of US$780 in 1980. The principal exports are copper (2.5 per cent of the world output in 1982) and gold (1.8 per cent of the world output in 1982) from Bougainville Island, the world's fourth largest copper mine. Another large copper and gold deposit is being developed in Ok Tedi.

Malaysia is the oldest among the developing countries surveyed, having obtained its independence in 1957, as well as the richest with a per capita income of US$1670 in 1980, and US$1740 in 1982. The mineral sector accounted for 4.6 per cent of the GDP in 1980 compared with 6.3 per cent in 1970. This decline was attributable to the over-all low rate of growth for the sector with the depletion of reserves in existing tin-bearing areas while the impact of the oil industry was felt only towards the second half of the decade. Value added in the sector grew at 4.6 per cent per annum in real terms. The growth during 1971–75 was 0.4 per cent per annum compared with 8.9 per cent during 1976–80. The higher growth rate during the latter part of the decade was due to the increase in production of crude oil. In 1970 the total export value of major minerals was 22.8 per cent of total merchandise export receipts, rising to 34.5 per cent in 1980, giving a growth rate of 23.6 per cent per annum. Total employment in the mining sector remained fairly stable, increasing slightly from 88 600 in 1970 to 89 600 in 1980.

Malaysia is the world's largest tin-producing country, accounting for 30 per cent of output in 1982. Other major hard rock minerals include copper (mainly from Sabah), bauxite (Johore), iron ore, kaolin and gold.

Work on the information of the national mining code was initiated in 1977 to establish the mining industry on a firmer basis through the standardisation of all mining enactments pertaining to prospecting, land alienation, issue and renewal of leases and conversion of mining land in the various states. A national mineral policy was also initiated in 1980 to encourage diversification beyond alluvial tin mining and to ensure orderly exploitation and development of non-hydrocarbon minerals. Several state governments are unhappy with the draft of the code and are negotiating directly with mining companies in matters connected with lease renewals and new concessions.

Quebec is a net exporter of metal and industrial minerals, and contributes 14.5 per cent of the value of Canada's metal production and 26 per cent of Canada's industrial mineral output. On an aggregate basis, in 1980 Quebec contributed 7.7 per cent of Canada's value of mineral production, compared to 10.4 per cent in 1974. Based on 1981 figures, iron ore formed 24.7 per cent of the total value of production, followed by asbestos (19.8 per cent), gold (12.5 per cent) and copper (8.4 per cent). In the case of iron ore, 75 per cent of the production leaves the province as concentrates or direct shipping ore, while for asbestos, only 2–3 per cent of the fibre is actually processed.

The key economic indicators are presented in Table 1.1. It can be seen that minerals form over 30 per cent of the exports of Quebec, and have an important 'macro' effect on the economy. At the regional level in many cases, the industry is not only the main employer but also the only reason for the existence of certain urban communities.

Although the discovery of gold was of particular significance in the early development of the **Western Australian** economy, renewed importance was due to the expansion of iron ore and other minerals in the late 1960s. This recent growth in importance of the industry is shown by the fact that in 1979–80 the value added by mining establishments was US$1.1 billion, or 510 per cent more than in 1968–69, when value added data first became available. In 1979–80 increases in gold prices gave a fillip to the industry, and the value of gold production in 1979–80 was exceeded only by that of iron ore and nickel. Other minerals mined include bauxite, coal, tin, tantalite-columbite and mineral sands.

Table 1.1 shows that although minerals form only about 2 per cent of the value of Western Australia's exports (this figure could be an underestimate since some of the produce could have been re-exported via other states), the industry provides over 20 per cent of the total employment, and as in Quebec, the sole *raison d'être* of certain communities.

TABLE 1.1 Key economic indicators for mining in selected countries (1980)

	Sierra Leone	Tanzania	Zambia	Botswana	Papua New Guinea	Malaysia	Quebec (Canada)	Western Australia
Value of product US$ (million)	125.8	13.0[a]	855.0[c]	167.0	na	528.0	2084	1107[b]
Contribution to Gross Domestic Product (GDP) (%)	11.5	4.0[b]	26.0[c]	31.6	15.0[a]	4.6	11.0[e]	5.5[e]
Contribution to total value of exports (%)	74.5	7.2	94.8[c]	69.5	23.5	23.6	30.6[c]	2.0[d]
Contribution to total government revenue (%)	1.0	na	26.7[a]	14.1	15.0[a]	na	na	na
Contribution to total employment (%)	8.1[c]	1.0	14.0[b]	8.8[a]	7.2[a]	1.7	1.2	20.3[d]

[a] 1978; [b] 1979; [c] 1981; [d] 1979–80; [e] This shows the contribution of mining to the whole country.

SOURCES

(i) Yearbook of Labour Statistics (ILO, Geneva, 1980).
(ii) Bank of Sierra Leone Economic Review (July–December 1980).
(iii) Quarterly Economic Review of Zambia, 4th quarter, 1982 (The Economist Intelligence Unit, London).
(iv) Republic of Botswana Employment Survey (August 1977).
(v) Bank of Tanzania Economic and Operations Report (June 1981).
(vi) Zambia, Monthly Digest of Statistics vol. 17, nos 4–6 (1981).
(vii) Botswana Statistical Bulletin vol. 7, no. 1 (March 1982).
(viii) Zambia Basic Economic Report (October 1977).
(ix) Mining Annual Review (1982).
(x) Western Australia Annual Year Book (1982).
(xi) Review, Canadian Mineral Industry (1981).
(xii) Mining Industry in Quebec (Department of Natural Resources, 1976).
(xiii) Fourth Malaysia Plan (1981–1985).
(xiv) Economic Report 1982–1983 (Ministry of Finance, Malaysia).

2 Organisation and Administration of Sector

MINERAL RIGHTS

The survey of mining legislation in the eight Commonwealth countries demonstrates a clear separation between surface rights and the rights to minerals whether they occur on the surface as alluvial or residual products or as sub-surface accumulations. The domanial system, contrasting to the traditional accession system of the old English common law, is the operative norm, since all the statutes have vested the rights to minerals in the state whether called the 'State' as in Tanzania, Sierra Leone, or Papua New Guinea, the 'President on behalf of the Republic' as in Zambia or the 'Crown' as in the case of Western Australia and Quebec (Canada). However, vestiges of the accessional system were retained for grants made before certain periods as demonstrated in Table 2.1. Even in the latter case, these rights are circumscribed by various provisions which in fact retain for the State the right to administer the method of development of the mineral resources. The African countries in the study retain ancestral rights to certain building and industrial minerals free of any impost, but retain for the State the right to supervise their disposal. In the case of Perak State, Malaysia, the ancestral rights to alluvial tin can be lost if the holders do not work the areas for two years.

ADMINISTRATION

Administration of the mineral sector is the responsibility of the Minister, who is assisted by a Chief Executive – the Director of Mines in Papua New Guinea and Western Australia, the Commissioner of Mines in Tanzania, the Chief Inspector of Mines in Sierra Leone, the Chief Mining Engineer in Zambia and Botswana, the Deputy Minister in Quebec (Canada). These functionaries are assisted in their duties by

TABLE 2.1 Summary of mining legislative provisions

	Sierra Leone	Tanzania	Zambia	Botswana	Papua New Guinea	Perak (Malaysia)	Quebec (Canada)	Western Australia
1. Mineral Rights								
(a) Domanial	Yes	Yes	Yes	Yes	Yes	Yes	Yes	Yes
(b) Accessional	Awards before 1927	No	No	No	No	No	Awards before 1880	Awards before 1899
(c) Ancestral	Yes. Over iron, salt, soda and potash. Not under any licence.	Yes. Over bldg materials and others.	Yes. Over bldg and industrial materials.	Yes. Over bldg and industrial materials.	–	Yes. Grants made before 1899[1]	–	–
(customary) rights								
(d) Vesting authority	Minister	Minister	Minister	Minister	Minister	Mentri Besar (Chief Minister)	– Minister	Minister
(e) Executive authority	Chief Inspector of Mines	Commissioner of Mines	Chief Mining Engineer	Chief Govt. Mining Engr.	Director	Chief Inspector of Mines (Collector)	Deputy Minister	Director of Mines
2. Licensing System								
A. Licenses issued[1]	PR;EPLi;MLe; Mr;AGMLi;DLiPR;	RLi;PLi;NLi; Cl.	PLi;Eli;MLi; Mp;PLibi;	Rp;PLi;Mle; RPLibi;RMLebi;	PA;Mle;GMle; DSLe;SMLe; LeMP	PLi;MC;Mle	PLi;Dli;Mle; EP;MC	PLi:Eli;MLe
B. Distinguishes between:			MLibi	MPbi.				
(i) Scale of operation	Yes	Yes	Yes	Yes	Yes	No	No	No
(ii) Mineral and mining type	Alluvial, lode Minerals	Minerals; Bldg materials; radioactive minerals	Minerals; Bldg and industrial minerals	Minerals; Bldg materials; radioactive minerals	Minerals; Gold; Dredging.	Tin; Minerals; Hydrocarbon	Alluvial Minerals; Hydrocarbon	Alluvial Minerals
(iii) Hydrocarbons included	No	No	No	No	No	No	Yes	No
(iv) Stages of development:								
(a) 1-tier	AGMLi;DL PR-MR;EPLi-MR	–	MP;MLibi	MPbi	MR	MLe	–	No
(b) 2-tier	PR-EPLi-MLe	PR-Cl RLi-PLi-MLi	PLibi-MLibi PLi-Eli-MLi	RPLibi-RMLebi PA-MLe RP-PLi-Mle	PA-MLe	Pi-MLe	–	
(c) 3-tier							PLi-DLi-Mle	PLi-ELi-Mle

C. Conditions I. Licence	Prospecting Rights	Reconnaissance Licence	Prospecting Licence	Reconnaissance Permit	Prospecting Authority	Prospecting Licence	Prospectors Licence [n]	Prospecting Licence
(a) Duration	1 year	1 year	4 years	1 year	2 years	Variable	1 year	2 years
(b) Renewal	Possible	Possible		Possible	Possible		Possible	Possible
1. Number	Indefinite	1	) No provision	Indefinite	Indefinite	) No provision	Indefinite	Indefinite
2. Duration	1 year each	1 year	)	1 year each	2 years each	)	1 year each	1 year each
3. Other conditions	same	same	)	May change	May change	)	Same	Same
(c) Size	Unspecified	Unspecified	) Unspecified	Unspecified	25 000 sq. km (m ax)	Unspecified	200 acres (surveyed) 225 acres (unsurveyed) per licence 180 000 acres per person	200 hectares (max.) per licence[p]
			) 13 sq. km for) industrial) minerals					
(d) Work programme	Not required	Required	Required	May be reqd	Required	Required	Not required	Required
(e) Financial guarantee for work programme	No	Yes	Yes	May be reqd	Yes	Yes	Not required	Required
(f) Relinquishment	To meet EPLi limit	No provision	To meet Eli limit	To meet PLi limit	50% of area currently held to be relinquished at each renewal	No provision	To meet DLi limit	To meet Eli limit
(g) Mineral disposal	Not possible	Not possible	Not possible	Not possible	Not possible	Possible	Not possible	Not possible
(h) Assignment	Not possible	With permission	With permission	Not possible	Not possible	Not possible	Not possible	Not possible
(i) Records and reports reqd.	No	Yes	Yes	Yes	Yes	Yes	No	Yes
(j) Exclusivity over:								
1. Mineral	No	Yes	Yes	No	Yes	Yes	Yes	No
2. Land	No	No	No	No	No	No	No	No
(k) Variation possible	Not necessary	Yes	Yes	Yes	Yes	Yes	Yes	Yes
(l) Rights acquired by:								
1. Citizens	Yes	Yes	Yes	Yes	Yes	Yes	Yes	Yes
2. Foreign indivs.	Yes	No	No	Conditional	–	Yes	Yes	Yes
3. Foreign Incorporated Cos	Yes	Yes	Yes	Yes	Yes	Yes	Yes	Yes
4. For Unincorp. Co.	Yes	With permission	No	With registration	No	Yes	Yes	Yes

	Sierra Leone	Tanzania	Zambia	Botswana	Papua New Guinea	Perak (Malaysia)	Quebec (Canada)	Western Australia
C. Conditions (contd)								
II. Licence	Exclusive Prospecting Licence	Prospecting Licence	Exploration Licence	Prospecting Licence	Prospecting Licence		Development Licence	Exploration Licence
(a) Duration	1 year	3 years	3 years	3 years	—	—	1 year	5 years
(b) Renewal								
1. Number	(5 for lode op. 2" alluv. op.	2	1, more for exceptional case	2	—	—	Indefinite	Indefinite
2. Duration	1 year each	2 years each	2 years	2 years each	—	—	1 year each	1 year each
3. Other conditions	May change	May change	May change	May change	—	—	Same	same
(c) Size	8 sq.ml (max.) 2 ″ ″ for PMi	Unspecified	26 sq.km (max.)	1000 sq.km (max.) 10 sq.km for indust. minerals	—	—	225 acres (max.)	10 sq. km 200 sq.km
(d) Work programme	Required	Required	Required	Required	—	—	Required	Required
(e) Financial guarantees for Work programme	May be reqd	Required	Required	Required	—	—	Required	Required
(f) Relinquishment	Not required	Reduction by half on each renewal	Not required	At end of licence or any renewal, area held to be halped each time.	—	—	Not required	After 3 yrs; After 4 yrs, half of remainder
(g) Mineral disposal	Not possible	Not possible	Not possible	Not possible	—	—	Not possible	Not possible
(h) Assignment possible	With consent	With consent	With Consent	With consent	—	—	Yes without consent	With consent
(i) Records and reports	Required	Required	Required	Required	—	—	Required	Required
(j) Exclusivity over								
1. Mineral	Yes	Yes	Yes	Yes	—	—	Yes	Yes
2. Land	No	No	No	No	—	—	No	No
(k) Variation possible	Yes, by Minister	Yes, by Minister	—	Yes, by Minister	—	—	Yes	Yes
(l) Rights acquired by:								
1. Citizens	Yes	Yes	Yes	Yes	—	—	Yes	Yes
2. Foreign indivs.	Yes	No	No	No	—	—	Yes	Yes q
3. For. Incorp. Co.	Yes	Yes	Yes	Yes	—	—	Yes	Yes q
4. For. Unincorp. Co.	Yes	With permission	Yes	Yes, with registration	—	—	Yes	Yes q

III. Licence	Mining Lease (MLe)	Mining Licence (MLi)	Mining Licence	Mining Lease	Mining Lease	Mining Lease	Mining Lease	Mining Lease
(a) Duration	5 to 99 yr	25 years	(25 yr (max.) (15 yr for indus. mins.	(25 yr (max.) (15 yr for indus. mins.	21 yr (max.)	21 yr (max.)	5 to 20 yr	21 yr
(b) Renewal								
1. Number	Possible Once	Possible Once	Possible Once	Possible Once	Possible Once	Possible	Possible 3 times	Possible Indefinite
2. Duration	Up to 99 yr	15 yr	(25 yr (max.) (15 yr for indus. mins	(25 yr (max.) (12 yr for indus. mins	21 yr	No provision	10 yr each	21 yr each
3. Other conditions	May change	May change	May change	May change	May change	—	May change	May change
(c) Size	Unspecified	Unspecified	Unspecified	Unspecified	(100 hectares (max.) (20 hectares (for gold	Unspecified	225 acres	10 sq. km
(d) Work programme	Required	Required	Required	Required	Required	Required	Required	Required
(e) Financial guarantees for work programme	Not read	Not reqd	Not reqd	Not reqd	Required	Required	Not reqd	Not reqd
(f) Relinquishment	Not necessary	Not necessary	Not necessary	Not necessary	Not necessary	Not necessary	Not necessary	Not necessary
(g) Mineral disposal	Possible	Possible	Possible	Possible	Possible	Possible	Possible	Possible
(h) Assignment	With permission	With permission	With permission	With permission	With permission	With permission	Yes, without permission	With permission
(i) Records and reports	Required	Required	Required	Required	Required	Required	Required	Required
(j) Exclusivity over:								
1. Mineral	Yes	Yes	Yes	Yes	Yes	Yes	Yes	Yes
2. Land	Yes	Yes	Yes	Yes	No[h]	Yes	No	No
(k) Variation possible	Yes, by Minister	Yes, by Minister	Yes, by Minister	Yes, by Minister	Yes[j]	Yes	Yes	Yes
(l) Rights acquired by:								
1. Citizens	Yes	Yes	Yes	Yes	Yes	Yes	Yes	Yes
2. Foreign indivs.	Yes	No	No	No	Yes	Yes	Yes	Yes
3. For. Incorp. Co.	Yes	Yes	No	With registration, yes	Yes	Yes	Yes	Yes
4. For Unincorp. Co.	Yes	With permission			Yes	Yes	Yes	Yes

	Sierra Leone	Tanzania	Zambia	Botswana	Papua New Guinea	Perak (Malaysia)	Quebec (Canada)	Western Australia
C. Conditions (contd)								
IV. Licence	Mining Right (MR) (for alluvs. only)	Prospecting Right (in designated area	Mineral Permit	Bld & Indust. Minerals Permit	Miners Right (essentially for alluvials)	—	—	—
(a) Duration	1 year	1 year	1 year	5 years	10 years (max.)	—	—	—
(b) Renewal	Yes	Yes	Yes	Yes				
1. Number	Indefinite	1	Indefinite	1	—	—	—	—
2. Duration	1 year each	1 year	1 year each	5 years	—	—	—	—
3. Other conditions	May change	Same	May change	May change	May change	—	—	—
(c) Size	880 yd along stream	Indefinite	2 hectares (max.)	0.5sq. km (max.) j		—	—	—
(d) Work programme	Not required	Not required	Required	Required	Not required	—	—	—
(e) Financial Guarantees								
1. For Work programme	No	Not required	Not required	Not required	Not required	—	—	—
2. For compensation to others	May be reqd.	Not required	Not required	Not required	Not required	—	—	—
(f) Relinquishment	Not required	Not required	Not required	Not required	Not required	—	—	—
(g) Mineral disposal	Yes	Yes	Yes	Yes	Yes	—	—	—
(h) Assignment reports	No	No	No	No	Yes	—	—	—
(j) Exclusivity over:								
1. Mineral	Yes	No	Yes	Yes	Yes	—	—	—
2. Land	No	No	No	No	No	—	—	—
(k) Variation possible	Yes, by Minister	No	No	Yes	Yes	—	—	—
(l) Rights acquired by								
1. Citizens	Yes	Yes	Yes	(Only for citizens (whether individual or company) RPLib^e	Yes	—	—	—
2. Foreign indivs.	Yes	No	No	RPLib;RMLeb^e	Yes	—	—	—
3. For. Incorp. Co.	Yes	No	No	GML;DSLe;LeMp^f	Yes	Permit^o	—	—
V. Other Licence	AGMLi DL^a	Claims^c	PLibi;MLibi^e		GMLc. DSLe;		Exploration	MR^c, GPL^e
V. Other Licence	AGMLi;DL^c	Claims ^{c}	PLibi; MLibi^e		GMLc.DSLc;		Exploration	MR^{c}GPL^e

3. *Local Participation*[d]						
(a) Exploration stage	Not necessary	Not necessary	(Only local (citizens	Not necessary	(Desirable[m]	Not necessary	Not necessary
(b) Development and mining stage	Desirable	Essential		Desirable	(Desirable	Not necessary	Desirable
4. *Dispute Settlement*						
(a) Executive decision	In some cases	In some cases	In some cases	In some cases	In some cases	In some cases	In some cases
(b) Courts and arbitration	Frequently available	Frequently available[d]	Frequently available	Frequently available	Frequently available	Always available	Always available
5. *Special Agreements*						
(a) Possible	Yes	Yes	Yes	Yes	Yes	Yes	Yes
(b) Parliamentary ratification necessary	Yes	No	No	No	No	No	No

Abbreviations used:

AGMLi	Alluvial Gold Mining Licence
CL	Claim
DLi	Development Licence
DSLe	Dredging or Sluicing Lease
ELi	Exploration Licence
EP	Exploration Permit
EPLi	Exclusive Prospecting Licence
GMLe	General Mining Lease
GPLe	General Purpose Lease
LeMP	Lease for Mining Purposes
MC	Mining Concession
MLe	Mining Lease
MLi	Mining Licence
MLibi	Mining Licence for building and industrial minerals
MP	Mining Permit
MPbi	Mining Permit for building and industrial minerals
MR	Mining Right
MrR	Miner's Right
PA	Prospecting Authority
PLi	Prospecting Licence
PLibi	Prospecting Licence for building and industrial minerals
PR	Prospecting Right
RLi	Reconnaissance Licence
RMLebi	Restricted Mining Lease for building and industrial minerals
RP	Reconnaissance Permit
RPLibi	Restricted Prospecting Licence for building and industrial minerals
SMLe	Special Mining Lease

Notes to Table 2.1

a An Alluvial Gold Mining Licence is issued to local residents to mine gold in a licensed mining area which is designated by the Minister. No more than five assistants can be employed.

b A Dredging Licence is required for carrying out dredging operations under conditions as the Minister may see fit. Local participation is categorised as 'desirable' when it is only covered by policy pronouncements. It is considered as 'essential' when it is enshrined in the statutes.

c A Prospecting Right is issued to a Tanzanian individual or company for rights to prospect in 'Designated Areas' for specified minerals which may be sought for and developed with a small amount of capital. It is issued for one year with the right of renewal for one further year to enable the holder to locate and peg mining claims which are renewable annually as long as mining takes place. Size of claims depends on mineral sought.

d Any decision of the Commissioner of Mines, except in relation to fees, can be contested in court. The Minister's decision in any appeal of the Commissioner's decision made to him with respect to registering and award of claims is final.

e Prospecting and Mining Licences for building and industrial minerals are only issued to citizens of Tanzania whether individually or collectively in a company. The conditions of application and grant along with the obligations and rights of the applicants are similar to those for a prospecting or mining licence for other minerals with notable exceptions with respect to duration and size.

f The Restricted Prospecting and Mining Licences for building and industrial minerals are only issued to citizens of Tanzania. Other comments in case of Tanzania at e also apply.

g Other prospecting authorities can be issued for different minerals for the same area, but priority is accorded to the first applicant.

h Holder of a Mineral Lease can acquire exclusivity to surface area by applying for a Lease for mining purposes.

i The major variation occurs in the grant of a Special Mining Lease (SMLe) which embraces the scenario of a development which is too large for the grant of a Mineral Lease or a Gold Mining Lease in the case of gold. The Special Mining Lease is generally for an area of 60 sq. km and is for 42 years with the right of renewal for 21 years under conditions which the Head of State sees fit. The SMLe is only issued by the Head of State acting on the advice of the Mining Advisory Board.

j A Miner's Right entitles the holder to locate an indefinite number of claims (see also 'r' below).

k A Gold Mining lease (GMLe) is issued on the same terms and conditions except for size as a Mineral Lease (MLe) for other minerals. It is issued for the exploitation of gold.

l A Dredging or Sluicing Lease (DSLe) is issued under similar conditions to a Mineral Lease but restricts the applicant to use that form of mining. A Lease for Mining Purposes (LeMP) is issued to the holder of other exploitation leases to occupy the surface land to perform the various works, for example, construction, and so on, which may be required. It runs for the duration of the exploitation lease in question. These rights can only be exercised in respect to alluvial tin deposits and can be forfeited if work is not carried on for two consecutive years.

m Significant fiscal incentives are available for companies meeting certain localisation guidelines (see Appendix 1(F)).

n A Prospector's Licence allows one to stake five claims for a total size of 200 acres in unsurveyed territory and 225 acres in surveyed territory. These claims are valid for 12 months, or 24 months if they are located north of latitude 52° N. The claims staked under a Prospector's Licence remain in force as long as the authorisation viz. Development Licence, Mining Licence, granting them are in force. A claim not so preserved cannot be restaked within 60 days by the holder of the Prospecting Licence.

o The Lieutenant-Governor in Council may make regulations authorising the Minister to issue Exploration Permits (EP) with varying conditions for exploration in New Quebec. The size of area under an EP is less than 25, but greater than 150 sq. miles, and the rights are valid for 10 years. In the case of alluvial minerals, EP can be granted for the entire province, but the minimum area is one sq. mile. The holder of an EP may apply for and obtain a mining lease for not more than 1/10th the area of the permit.

p One person can hold up to ten Prospecting Licences without the approval of the Minister. Additional amounts can be obtained only with permission.

q Australian foreign investment rules require local participation at this stage. In fact, the government guidelines indicate that a proposed project for the production of uranium must have 75 per cent Australian equity interest, and will only be allowed to proceed if this interest is not available locally, and that the project at least has a 50 per cent Australian equity, and control of the management. This minimum equity position must be increased over time. For other mining projects which involve a total investment greater than A$5 million, approval will not be given except a 50 per cent equity and voting strength resides with Australians.

r A Miner's Right (MR) permits the holder to prospect for and locate claims on a non-exclusive basis.

A General Purpose Lease (GPLe) allows the holder of mineral rights to use the surface to implement works which are essential for mining.

Directors of Geology as in the case of Zambia, in matters needing geological inputs, or Wardens as in Papua New Guinea on matters of 'policing' of Act and its regulations. These chief executives have quasi judicial roles since the Act confers on them the ability to receive evidence and render decisions in cases of dispute.

THE LICENSING SYSTEM

The licensing system of the eight countries is basically designed around three concerns, namely:

(i) The scale of contemplated mining development.
(ii) The mineral type and the method of mining.
(iii) The stages of mineral development envisaged.

The scale of mining development, though not defined in absolute terms in any of the mining acts surveyed, implicitly affects the type of licences awarded and the conditions of their grant. It is only in the context of taxation, and royalty payments as in the case of Tanzania's diamond levy, the royalty provisions on diamonds in Botswana, or in the definition of size for Australia's localisation requirements, that size is absolutely defined. In the context of the mining acts as typified by Article 69 of the Tanzanian Code, the size is explicitly recognised in the grant of claims in cases where 'the Minister considers that it would be in the public interest to encourage prospecting and mining for minerals in any area of land by methods not involving substantial expenditure or the use of specialist technology'. Similar concerns are embodied in the grant of Mining Permits to mine building and industrial minerals as in the case of the Zambian Act (Art. 69).

The licences which are awarded for relatively unsophisticated, and essentially more labour-intensive operations, generally involve smaller areas, while the complex of conditions which are required for the larger developments are considerably simplified and reduced. Invariably, the public functionaries charged with the responsibility for the administration of these licences have wider discretionary powers as opposed to the case of larger developments. Awards of licences of this type are invariably restricted to local citizens, viz. Zambia's Art. 64, Tanzania's Art. 70 and Botswana's Art. 53. Neither the older legislation of Sierra Leone nor the legislation of the two developed countries contains provisions of this nature.

The type of mineral and the methods used for its extraction have profound effects on the licensing system of all the countries considered. All the acts contain definitions of minerals which establish the basis for the licensing system. For example, Quebec recognises 'minerals' or 'mineral substances' as 'all natural solid, liquid or gaseous mineral substances, and all fossilised organic matter'. By so defining 'mineral', the Act establishes the framework for the administration of the search for, development, and disposal of, all hard minerals and hydrocarbons. Perak State of Malaysia followed this trend up to 1974 when petroleum development was made the responsibility of PETRONAS – a state oil corporation which manages all stages of the development of petroleum.

The newer statutes of Botswana (1976), Zambia (1976), Papua New Guinea (1978), and Tanzania (1979) have all sought to concentrate solely on the hard minerals, while leaving hydrocarbons the subject of separate and distinct statutes. No doubt, the special nature of petroleum in terms of its tremendous impact on the economies of under-developed countries in the post–1973 period, coupled with its huge financial and technological inputs, has inspired this trend. The Tanzanian definition of mineral and its exclusion of petroleum is instructive. A mineral is defined as 'any substance, whether in solid, liquid or gaseous form, occurring naturally in or on the earth, or in or under the seabed, formed by or subject to a geological process, but does not include mineral oil (as defined in the Mining/Mineral Oil Ordinance) or water'. The definitions are similar in the case of Botswana, Zambia and Papua New Guinea.

Notwithstanding the all-encompassing definition of minerals, the peculiar endowment and practice of specific states require that still further categorisation be made. For purposes of isolating specific mineral products for separate treatment, hence for licensing, the mining statutes of the various countries have further reflected this fact. In the Acts of Tanzania, Zambia and Botswana, building and industrial minerals are identified for different licensing arrangements while in the case of Sierra Leone, Tanzania, Perak (Malaysia), and Quebec (Canada) alluvial minerals including gold and precious stones have been isolated for special treatment (see notes on Table 2.1 for the various specific mineral or mining licences recognised). Radioactive minerals are explicitly treated in the case of Tanzania (Art. 67) and Botswana (Art. 45), reflecting those states' concern with respect to the disposal and use of these products. Zambia isolates 'reserved minerals' for special attention, and hence concessions within its licensing system, while Tanzania refers to 'specified minerals'. The need to further dissect the definition of mineral becomes even more apparent for the purposes of royalty payments as can be seen in Appendix 1 (at the end of this book).

All the licensing systems explicitly take into account the various stages of the mining cycle viz.:

1. The search for an anomalous mineral occurrence.
2. The geological definition and economic assessment of the mineral deposit.
3. The preparation (development), and extraction of the ore deposit.

Six of the eight countries issue licences for each of these stages. The various names given to the licences are identified in Table 2.1, and briefly summarised they are:

Stage 1 – Prospecting Right (PR); Reconnaissance Licence (RLi); Prospecting Licence (PLi); Reconnaissance Permit (RP).
Stage 2 – Exclusive Prospecting Licence (EPLi); Prospecting Licence (PLi); Development Licence (DLi); Exploration Licence (ELi).
Stage 3 – Mining Lease (MLe); Mining Licence (MLi).

Generally, the three-tier system of licensing is reserved for major mineral developments involving huge deployment of capital and specialised skills, though in the case of Quebec (Canada) and Western Australia it represents the framework for development of all minerals. The other countries (Sierra Leone, Tanzania, Zambia and Botswana) have created a simpler two-tier system to regulate the exploration for, and development of, smaller and less sophisticated mineral projects involving alluvial minerals and building and industrial products.

Perak (Malaysia) and Papua New Guinea have integrated the exploration and development stages and therefore have two-tier systems of licensing though, in the case of Papua New Guinea, there is provision for the issuing of a special mining lease for projects of an extraordinarily large size. This has no doubt been conditioned by their experience with the huge Bouganville, and Ok Tedi porphyry copper projects. Sierra Leone, Zambia, Botswana, Papua New Guinea, and Perak (Malaysia) also have one-tier licensing systems which allow for extraction without any preliminary exploration of already defined, surficial deposits of alluvial minerals or building and industrial products.

Conditions of Licence for Searching for Mineral Deposits

The applicant for a licence to search for mineral deposits is required to show, via his application and further representations, that he is financially and technically capable of carrying out a programme of prospection which he has prepared for the mineral (s) he has identified, at the level of expenditure contemplated. He may be further required to submit additional information on himself or his company – if he represents one – or any other detail which the Minister may require. When considering his application the Minister may impose other conditions, which he considers necessary for the grant of the licence. The Minister would normally reject the application if it covers an area held under a mining tenement for the same mineral, or if the applicant is not technically nor financially capable, nor is eligible under the Mining Act (see Art. 18, 19 of Zambia, 19 to 21 of Tanzania, 27 of Papua New Guinea, 14 of Botswana, 41 of Western Australia).

Once the grant is made it is generally for a short period, usually one to two years (see Table 2.1), and can be renewed for further periods at the discretion of the Minister, who must be satisfied that the lands were beneficially occupied and the work programme was vigorously implemented. To ensure compliance with the terms of the licence, particularly with respect to the accomplishment of the work programme, the applicant is required to guarantee by way of a bond or similar device to undertake minimum expenditure requirements which can be recovered by the State for non-fulfilment of the work programme objectives. In Zambia, work equivalent to an expenditure of 10 kwachas per sq. km (k^2) must be done (Art. 26). In Perak (Malaysia) a security of M$1000 may be required, while in Quebec it ranges from C$2 to C$6 per acre depending on location and the age of the licence.

In the Quebec system, the Prospector's Licence entitles the holder who, through a 1977 amendment to the Mining Act, must achieve ministerial approval, to peg at least five claims of 225 acres for each licence he holds. The licence is valid for one year, and he is entitled to as many licences which would allow him to peg an aggregate of 180 000 acres to 600 000 acres depending on the location.

Generally, the sizes of areas granted under licences at this stage are quite large. Table 2.1 also demonstrates that in a number of cases, no limits are set on the size of area. Relinquishment of area may be required during the tenure of this licence, but is definitely needed to meet the size specifications of the next licensing stage which invariably always sets size limits. A form of inducement for relinquishment is also to be found in annual escalating area rentals (see Appendix 1(B) and 1(D)).

It is not usually possible for licences of this type to be freely assigned to some other individual, and while it is possible in the case of Tanzania (Art. 58) and Zambia (Art. 39), it must be accomplished by way of ministerial approval. Note, however, that in the case of Quebec (Art. 195), inasmuch as the Prospector's Licence cannot be assigned, the claims located under those licences can be freely dealt with, but there is an obligation for 'post facto' registration.

In Sierra Leone, Botswana and Western Australia, the holders of licence at this stage, in effect, have non-exclusive rights over the area in question. The licences do not confer any exclusive rights of search over a specific mineral, since other applicants can be given rights to look for the same mineral over those areas. Tanzania, Zambia, Perak (Malaysia), Papua New Guinea and Quebec award exclusivity over the mineral to the applicant for the duration of the licence, or to the time of relinquishment of those rights, whichever occurs first. However, in this latter case, except Quebec, other licences can be issued for other minerals over the same area. These licences do not convey, with the exception of Malaysia, the right to remove any minerals from the area. However, provisions permitting the collection and disposal of samples from the area are generally present.

The maintenance of proper reports and records is required in most of the countries, and even when they are not statutorily required, as in the case of Sierra Leone and Perak (Malaysia), they are essential prerequisites in the considerations for a licence at the next stage.

Conditions of Licence for Geologically Defining and Assessing the Economic Potential of a Mineral Deposit

The holders of Stage 1 licences, who have successfully completed the requirements of that licence and have identified a mineral deposit, can in general terms apply for and be granted a licence to geologically define and economically assess the said deposit. However, this grant is never automatic, but is subject to ministerial approval which will be influenced by the nature of the work programme, the applicant's financial and technical capability, and a host of other considerations. All the statutes treat succession in licences in this way. If an applicant fails to obtain an award because of his inability to pass the qualification criteria, he loses rights over the deposit without being entitled to compensation.

Holders of Stage 1 licences are generally restricted in the use of definition surveys, for example, drilling, and are asked to constrain themselves to the use of techniques which give fairly wide coverage, for example, regional geochemistry, and remote sensing using imagery, photographs, or applied geophysics. At this stage, drilling and close-spaced geochemistry and geophysics are allowed.

The more detailed aspect of the work at this stage presupposes that smaller areas will be required and Table 2.1 shows the range of sizes which are current in the respective countries. The sizes, of course, reflect to some extent the size of the country in question, the degree of competition and perhaps also the amount of available capital.

Quite often, as in the case of Tanzania, Botswana and Western Australia, specified relinquishment provisions are in force. The relinquishment provisions deal not only with the size of area, but also with the shape, since the State tries to reduce attempts at 'raisin picking' by the licence holder. Even where periodic relinquishment is not required, as in Papua New Guinea and Quebec, relinquishment is required to meet the requirements of even smaller sizes at the next stage. Relinquishment is probably also accomplished by the use of the escalating value of minimum work obligations, and escalating area rentals as demonstrated in Art. 143 of Quebec.

The licence holder is invariably required to guarantee either his entire work programme or some minimum work obligations. Art. 73 of Botswana is instructive of the kind of guarantees or 'security of compliance' as it is called. It states that the Minister may, from time to time, make such arrangements as to him appear appropriate to secure that the holders of mineral concessions comply with the provisions of this Act, and without prejudice to the generality of the foregoing, may accept guarantees, whether from shareholders or otherwise, in respect of such compliance. The Tanzanian statute (Art. 36) goes even further by stating that any amount required to be expended on prospecting operations which is not expended is a debt due to the United Republic and is recoverable in a court of competent jurisdiction. Zambia (Art. 43) follows the same scheme.

Because work at this stage is generally a time-consuming and tedious process, the initial duration of these licences is generally longer than in Stage 1. Table 2.1 demonstrates a preference for three years. Once work progresses satisfactorily, renewals are invariably granted to allow the licence holder to define as precisely as possible the deposit he is investigating. However, the maximum period for these licences seems generally to be in the region of five to seven years. Conditions of renewal may change to conform with any statutory changes which may have been made or as the Minister may consider necessary, though invariably only with respect to work programme requirements (Art. 13 of Sierra Leone, 49 of Tanzania, 31 of Zambia, 20 of Botswana, 25 of Papua New

Guinea). Note, however, that in the case of Quebec, the Act (Art. 76 to 88) defines in very great detail the nature and amount of the required work, thus reducing the area for arbitrary discretion on the part of the Minister.

Large amounts of capital are generally expended in this phase of the mining cycle, and security of tenure, coupled with the right to freely conduct one's work, is at a premium. All the statutes preserve the right of the licence holder to exclusivity over the mineral in the area covered by the licence. While Zambia (Art. 31) and Tanzania (Art. 29) will not grant any other licences over the same area for any other mineral, Botswana (Art. 19) makes provision for the possibility of granting rights over other minerals in the same area.

Claims staked in Quebec on public lands carry with them exclusivity over minerals except hydrocarbons in that area with saving provisions allowing right of way to others, rights of the State to building materials, and the expropriation without compensation for public purposes. Claims held over private lands do not convey the rights to do any works without the authorisation of the owners of the land or by expropriation (Art. 221).

Assignment of mineral rights at this stage is possible with the prior approval of the Minister in all countries except Quebec where assignment of any interest in claims can be done without ministerial consent. However, the assignment must be registered. While the statutes dealing with assignment are quite general, Botswana (Art. 48) identified in particular the inability of anyone to assign to any group or person a financial interest exceeding 20 per cent which would carry a concomitant voting right, without ministerial approval.

Records and reports are essential at this stage, since the results of the work could lead to the establishment of a mine. The enabling provision could be as simply written as is done in the case of Botswana (Art. 22) and Zambia (Art. 37). It states that the licence holder is required to submit to such persons at such intervals such reports and such affidavits containing such information and supported in such manner as may be prescribed, or it could be more elaborate as in the case of Papua New Guinea where different kinds of quarterly, half-yearly and annual reports on physical work and financial transactions are required.

The most important report to be submitted at the end of this stage is the feasibility study on the technical and economic parameters which would establish whether a mine could be commercially established and under what conditions. The provisions of Art. 32 of Botswana are recorded as Appendix 2 to demonstrate the range of issues which must be contemplated by the feasibility study.

Conditions of Licence for the Development and Extraction of the Ore Deposit

Once the holders of Stage 2 licences in a three-tier system or Stage 1 licences in the case of a two-tier system as in the case of Papua New Guinea or Perak (Malaysia) have submitted their feasibility reports and have agreed with the Minister on the commercial viability of the deposit, along with the conditions under which mining will be permitted to occur, a licence to develop and mine will be granted. It should be observed that notwithstanding the fact that holders of Stage 2 licence in the three-tier system, or Stage 1 licence in the two-tier system have completed satisfactory work in establishing a commercial deposit, no automatic rights exist to a licencee to develop and mine a deposit. Rights to exploit can only be achieved on the applicants' qualifying on the myriad of issues, inter alia, the proposed programme, technical capability, financing availability and method, development concerns including infrastructure, local participation, and so on. This is a universal feature of all the countries studied.

The huge magnitude of capital required to establish a mine, and the potentially long periods which may be involved in its recovery, are generally reflected in licences at this stage. They are being issued for longer periods compared to licences in the former stages. Table 2.1 shows that 20 to 25 years is a typical period for initial grants, with an excessively generous maximum period of up to 99 years in Sierra Leone. Where industrial minerals are involved, the period is generally shorter as seen in Zambia and Botswana where 15 years is normal for initial grants. Renewal of these licences is always possible, but the conditions under which the licences were granted may be changed at each renewal. The provisions with respect to renewal in Art. 104 of Quebec, 45 of Papua New Guinea, 38 of Botswana and 52 of Tanzania are typical of the way in which renewals are handled. The Papua New Guinea Act (Art. 45) grants the right of renewal on such terms 'as are in force with regard to mineral leases at the time of renewal'. In Tanzania (Art. 52) the Minister 'shall renew the licence, with or without variation of the conditions of the licence'.

The size of area granted for mineral leases is not specified in five of the countries, while Quebec (Canada), Papua New Guinea and Western Australia do so. Where specified, the maximum area is less than that granted under Stage 2 licences. No relinquishment of area is statutorily required at this stage.

Holders of mining licences are granted exclusive rights to the mineral and preferential rights in the use of the surface to construct facilities required for mining. While the rights to the use of the surface for works ancillary to mining are automatic in Sierra Leone (Art. 35), Tanzania (Art. 43), Zambia (Art. 53) and Botswana (Art. 40), in Papua New Guinea (Art. 112) and Western Australia (Div. 4) these rights are conferred by the grant of Special Mining Easements and a General Purpose Lease respectively. Extracted minerals are the property of the licence holders and they could dispose of them freely. The Acts may, however, stipulate, especially in the case of gold and precious stones, that the minerals could only be sold to licensed buyers.

At the mining stage, the operation becomes a commercial one, and owners of licences are required to keep records of all their transactions, making periodic reports to the Minister.

Other Licences

Section 2C IV of Table 2.1 identifies the conditions which are generally applicable to mining for alluvials and building products, which do not require large outlays of capital. The licences are generally for shorter periods, and involving smaller tracts of land, while requiring fewer requirements for their award. The notes to Section 2C V of Table 2.1 further identify the other kinds of licences awarded.

Of particular interest is the award of a Prospecting Right in a 'designated area' by the Commissioner of Mines in Tanzania to search for 'prescribed minerals'. This is essentially the way in which small scale mining ventures are conducted as alluded to in an earlier section. The Prospecting Right confers the right to locate claims, which could then be worked, and the minerals taken possession of. If one were to be able to categorise licences in terms of exclusivity and the intertemporal definiteness of property rights, it is clear that this licence confers the least secure position. It cannot be issued in any area where a reconnaissance prospecting or mining licence has been granted, and must be confined to the 'designated area', while the mere act of annulling the order creating a 'designated area' renders the claims invalid. In Sierra Leone, successful prospection will lead to the grant of a Mining Right which conveys over that designated area, all the rights of a Mining Licence described in the last section. However, this Mining Right must be renewed annually.

The Sierra Leone Act (Art. 28) provides for the Minister requesting a holder of a Mining Right to apply for a Mining Lease, if the deposit is considered to be of the size and characteristics which would permit it to be more properly developed under the tighter regime of a Mining Lease.

One of the principal features of the 1976 amendment to the Zambian Mining Act (Art. 57) was to enable the Minister to designate an area 'a mine'. An area, once so designated, has to be developed within six months, and under conditions applicable to licences to mine. The net effect of these two provisions is to facilitate the timely and efficient development of the minerals from the standpoint of the state officials.

Industrial and building minerals are generally treated at two levels in the various mining acts. In one case, there is the customary or traditional rights approach where individuals or communities at different levels of organisation had rights to industrial minerals, which could be extracted with or without the issue of a permit. Materials obtained in this ways were not to be traded. In all the countries, mining rights also conferred rights to building materials in the area covered by the licence. Licences issued in Zambia and Botswana at this level are called 'Minerals Permit' and 'Restricted Minerals Permit' respectively.

In the case where building and industrial products were being prospected and exploited for the purposes of trading, a licensing system akin to that for large scale mineral development becomes applicable, with the basic differences that area sizes are smaller and duration of licence is shorter, as has been observed above. The provisions in Zambia and Bostwana are instructive. Zambia issues a Prospecting and then a Mining Licence for building and industrial minerals, while Botswana issues Restricted Prospecting and Restricted Mining Licences for industrial minerals. Table 2.1 demonstrates the similarities or differences in licence types of this nature and others, and one fact which should be recognised is that all the licences to building and industrial minerals can only be acquired by local citizens, except in Quebec (Canada) and Western Australia, where a more open policy to the acquistion of mineral rights is in practice.

Licences to mine are in some cases further categorised depending on the mineral or the mining type. Generally, the terms of the licences are very similar to the general cases, excepting perhaps a few conditions with respect to some specific detail. Licence to carry out underground mining may have special safety regulations with respect to stopes, and so on, as opposed to those conferring rights to surface mining where stopes are not involved. Quebec confers an 'underground mining lease' and a 'mining lease'; Sierra Leone grants an Alluvial Gold Mining Licence; Papua New Guinea grants a Dredging and Sluicing Lease, and the others, Zambia and Botswana, as indicated above for industrial minerals.

MINERAL RIGHTS AND SURFACE RIGHTS

At the opening of this section, it was stated that all the countries made a clear distinction between surface and mineral rights. Inferentially, it was therefore possible for there to be in existence simultaneous awards of surface and mineral rights to different applicants. The situation could arrive when the rights exercised by either licence holder were complimentary, neutral or competitive in use. Situations of complimentarity or neutrality in use are relatively harmonious states and do not therefore require much attention, from the standpoint of litigation. However, situations of competition in use are disputations. They could lead to a breakdown of order, and hence their resolution must be provided for by a clear set of codes.

The mining statutes of all the countries recognised the rights of the land (surface) owner to the peaceful enjoyment of the land especially with respect to the right to graze stock and cultivate crops. Art. 80 of Tanzania is instructive – 'The lawful occupier of any land in a reconnaissance area, a prospecting area, a mining area, or a claim area, retains any right which he may have to graze stock or to cultivate the surface of the land except in so far as the grazing or cultivation interferes with reconnaissance, prospecting or mining operations in any such area'. However, thereafter, there are certain restrictions on surface owners' or mineral rights owners' abilities to do some things without prior consultation.

For example, the owners of mineral rights are expected to gain the prior approval of surface owners before carrying out any works on the areas; or, as in the case of Quebec (Art. 221), rights to carry out these acts can also be acquired by expropriation. Where the surface owner, without good reason, refuses to grant the mineral rights owner permission to do certain acts, the mining administrators can overrule the rights of the surface owner and permit the acts to be undertaken by the holder of the mineral rights (see, for example, Art. 83 of Zambia).

Where the exercise of rights by the mineral rights owner causes damage to, or restricts the surface owner to the use and enjoyment of, his rights, all the Acts make provision for the payment of compensation, which invariably includes the value of any enhancement caused by the exploration activity. Art. 81 of Tanzania is typical – Where the value of any land has been enhanced by reconnaissance, prospecting or mining operations, compensation payable pursuant to subsection 1 in respect of the land shall not exceed any amount which would be payable if the value had not been so enhanced'. Papua New Guinea and Western

Australia use a similar scheme with the exception that it conveys the rights to perform surface works by the grant of special mining easement or general mining leases, as was discussed above.

NATIONALITY OF MINERAL RIGHTS HOLDERS AND LOCAL PARTICIPATION

The Quebec Mining Act (Sec.4) states that 'aliens, as well as Canadian citizens, may enjoy the benefits of this Act by complying with its provisions'. This has been the approach of the old mining statutes of the underdeveloped countries of the British Commonwealth, which have all since enacted various modifications restricting the right of aliens to hold mineral rights. Even in the case of Quebec, Western Australia and Papua New Guinea, which have retained on the statutes this general open policy, the rights of aliens are restricted by the foreign investment and localisation guideliness referred to at the beginning of this chapter.

The African countries and Perak (Malaysia) restrict the development of small-scale operations to nationals only. Provisions such as Art. 53 of Botswana which states, inter alia, 'subject to the provision of subsection 2, a building and industrial minerals permit shall not be granted to a person who is not a citizen of Botswana', are typical. The savings clause in subsection 2 further defines citizens in relation to a company or corporation where the 'membership is composed exclusively of persons who are citizens of Botswana . . . whose directors are exclusively citizens of Botswana . . . which is controlled by individuals who are citizens of Botswana'. Tanzania and Zambia have similar conditions.

Where the developments involving larger-scale, mineral activities are concerned, Table 2.1 identifies the attitude of the respective countries. The African countries make certain distinctions at the various stages of development where the essential point is that from the standpoint of the company, only those which are incorporated in the country can be awarded mining leases. Unincorporated foreign companies are generally permitted to carry out exploration activities, but must be incorporated before being awarded a mining lease. Art. 5 of Botswana is typical. It asserts, inter alia, that no mineral concession shall be granted to or held by a company which 'has not established a domicilium citandi et executandi in Botswana; or unless in the case of a mining lease, such company is incorporated under the Companies Act'.

Tanzania (Art. 32), Zambia (Art. 20) and Botswana (Art. 35) have all made provisions for the possibility of the government or persons nominated by the same to acquire a right in any mineral venture. This feature of the Act therefore paves the way for the government to take an active interest in the mining projects. The actual amount of the participation is not stated, and is left to be settled by negotiations, though Zambia in particular has stated, by various policy pronouncements, that it will seek a 51 per cent participation interest. Art. 31 of Tanzania further requires that at the stage of an initial grant of prospecting rights, the foreign company must put up proposals for the training of Tanzanian nationals. These development issues become even more relevant in the various statutes, especially at the stage of the granting of mining licences.

TERMINATION OF RIGHTS AND DISPUTE SETTLEMENT

Definiteness of property rights and security of tenure are essential ingredients in an investor's decision to deploy resources in order to generate an income stream. The mining acts attempt to remove any indefiniteness or instability in property rights by establishing the norms under which rights will be awarded, and the obligations of the holders of those rights. However, acts of non-compliance or misinterpretation on the part of participants in the mining sector, coupled with a real difficulty to precisely specify all the norms relating to accomplishment of a set of obligations, will always lead to disputes. Dispute settlement is therefore an essential concern in all the countries.

The mining acts of all the countries have assigned wide discretionary powers to the Minister in the granting or rejection of licences, in the settlement of disputes between contending parties with respect to the preservation of their perceived rights under various licences, and in the termination of rights awarded by licences. Some of these powers have been delegated to the public functionaries such as the Commissioner of Mines in Tanzania, the Deputy Minister in Quebec or the Warden in Papua New Guinea.

The preceding sections have outlined the various conditions under which various licences will be awarded. It is to be noted that mention is made of 'satisfactory' work programme, 'adequate' financial and technical ability, and so on – all terms which lead to the utilisation of acts of imprecise judgement, and hence discretion. Because of their contestable nature, most of the Acts also subject the Minister's decision to appeal in the courts. In the Quebec Act (Art. 277), a mining judge is appointed who shall have 'jurisdiction over all matters within the competence of the Minister under this Act (a) by way of appeal in cases where an appeal lies; (b) upon reference by the Minister in every case where the Minister deems expedient'. In Tanzania (Art. 85), Zambia (Art. 102) and Papua New Guinea (Art. 160), decisions of the Commissioner, Engineer and Warden respectively can be appealed in the High Court.

There are, of course, some notable exceptions as in Perak (Malaysia) (Art. 22) where the Mentri Besar's (Chief Minister's) ruling on the forfeiture of a mining licence is final and cannot be contested in a court of law. There is also the case of compulsory acquisition of land for public purposes in Papua New Guinea (Art. 68), Sierra Leone (Art. 76), Zambia (Art. 83) and Botswana (Art. 64). In Zambia (Art. 56) the President can order a cessation or curtailment of production in the national interest, while in Tanzania (Art.48) the Minister's decision with respect to the restriction on exercise of some rights of Mineral Rights holders is final. In Papua New Guinea (Art. 134), where both parties agree to have the Warden settle a complaint, his decision is final. These are but few examples of final decisions by members of the executive arm.

Invariably, the public functionaries have greater authority to adjudicate and render final decisions in the case of small-scale activities as is the case with respect to claims in Tanzania and Papua New Guinea. At this level, questions of disputes between claim holders, the observance of boundary lines, and so on, fall almost exclusively within the final decision-making ability of the functionaires. However, for the larger developments where more resources are involved, the Minister is the principal decision-maker in the initial interpretation of the laws. There is no provision in any of the Acts for the use of any other laws but local laws in the settlement of disputes between the state and extranational groups, though perhaps this may be found in other statutes.

A novel approach to dispute settlement was attempted in Zambia by the creation of a Mining Affairs Appeal Tribunal. Articles 111 to 126 of the 1970 Act elaborated in detail the functions of this group, which was empowered to review and render decisions on appeals of the decisions of the Minister and the Engineer, on matters concerning the rejection of applications for licences of all sorts, their renewal or their transference, or the termination of licences. The decision of the Mining Affairs Tribunal was to be final, and could not be challenged in any court. The Tribunal was composed of a number of high-powered public and legal officials. However, one of the major amendments to the 1970 Act, apart from the granting of the power to the Minister to declare 'an area a mine', was the abolishing of the Mining Appeal Tribunal. Obviously, the experiment did not work, and the 1976 Act has simply followed the pattern of the newer legislations and assigned to the High Court of the land the right of interpretation of the statutes.

SUMMARY COMMENT

The review of the legislative framework of the mining sector through the provisions of their various mining acts, on average demonstrates a large degree of similarity of treatment of the omnibus issues which are relevant to the development of a mineral sector. Though the statutes are essentially regulatory instruments, it was possible to note an evolution in the principal foci of the various statutes depending on the time they were constructed.

Basically, the Perak (Malaysia) State Statutes (Cap 147 of 1962) retained the character of the traditional, open-concession 'policing' style statute as Walde (1983) calls them. Essentially, he describes these kinds of acts as governing 'the issue, administration and cancellation of mining titles; it organised the national mining authorities and their powers and provided for supervision of mining activities (accidents, inspection, registration, dispute settlement between competing mining claims, enforcement)'. Developmental issues concerning training of nationals, relevant economic linkages, and so on, do not find their way into this type of statute, but these are contained in a general policy document known as New Economic Policy, which covers all sectors, including mining. The Sierra Leone Statute (Cap 196) follows the traditional trend, though to a lesser extent.

Interestingly enough, the Quebec mining statutes are essentially regulatory in nature, with the difference being that there is greater detail on issues of required work obligations of the claim holders, while defining more precisely the method public functionaries would adopt in the discharge of their duties. Their discretionary powers are more restricted. Presumably, the relatively developed state of Quebec, economically and institutionally, has resulted in the decreasing necessity to have development concerns reflected in the mining acts. It also probably reflects the lesser importance accorded to government as an economic change agent.

The African countries, Tanzania, Zambia and Botswana, which all have fairly new statutes (Tanzania 1979, Zambia 1976, Botswana 1976), have incorporated many developmental concerns. Some issues which are particularly dealt with are the use of local materials, training possibilities for citizens, environmental protection, infrastructure development, processing, development of ancillary services and actual participation in the financial and administrative management of the project. Though, in fact, Sierra Leone has one of the older style statutes, the mineral development agreements which it signed in the post–1966 period have all addressed, to a greater extent, the developmental responsibilities of the mineral sector.

3 The Financial Regime

The previous section alluded to the fact that true expressions of sovereignty will only be achieved by the appropriateness of the fiscal and other measures which determine whether a deposit is commercially exploitable and the return to the investor and the State which would arise as a consequence of its exploitation. The financial regime is conceived here to mean the range of levies which the State makes on a mining enterprise, along with the accounting allowances which it provides in the computation of those levies. It also includes other non-financial instruments, like government participation, which affect the level of return to equity holders.

The full range of levies encompasses once for all fees for specific transactions, rentals for use of land, signature or production bonuses (more relevant for oil), royalty payments for use of resource, income and corporate taxes, surtaxes, levies, import and export taxes or duties. Because the actual quantum of some of these levies is negligible compared to the size of the income stream, it was decided to isolate only the instruments which had a major impact in the determination of the project economics and hence the returns to the State and investor. Tables 3.1 and 4.1 and Appendix 1 indicate the instruments that were considered in this study. They represent the available information from various tax and other statutes from the respective countries. However, it is to be recognised that these instruments are constantly being reviewed and revised. This study attempted to use the most up-to-date information that was available.

A word of caution needs to be issued in evaluating the information of this section, which deals specifically with the major impositions which an investor will face in the respective jurisdictions. There are many *discretionary* incentives which could or may be granted to any project, but these are only listed as being available. They are not assumed to be given in the context of the financial model developed in chapter 4. For example, in Sierra Leone many mineral developments have been granted tax holidays for different periods, while in Tanzania the application of an additional profits tax has been used. These have been on a case-by-case basis and have therefore not been included as a generalised condition.

POLICY IMPLICATIONS OF INSTRUMENTS

Objectives of Government Policy

An analysis of the impact of the respective regimes must be set against the major policy concerns of the governments and the instruments they have used to achieve these objectives. Whether the instrument is consistent with the policy objective is partially the enquiry of this section.

Basically five objectives have been identified in the mineral sector of the various countries. These are:

1. Maximising government revenues over-time.
2. Immediate revenue collection.
3. Allocation of investment.
4. Integration of the mineral sector in the total economy.
5. Conservation of the mineral resources.

Some or all of these objectives, which may be inherently conflicting, find their way either into policy pronouncements of various governments or they may even be expressed in statutes in a tangible way, in terms of the choice of fiscal instruments which the state uses to regulate the sector. In Papua New Guinea one of the express goals of the government's mineral policy is to 'maximize revenue flows' from the exploitation of its mineral resources over time. There are, of course, other policy concerns, for example, the need for conservation and control of the activities of major enterprises in the exploitation of natural resources, but the revenue-maximisation concern will be reflected, if given physical expression, in the instruments which are used to obtain the government's portion of the revenues flowing from minerals projects. If revenue maximisation is of central concern, then the financial regime will be concessionary to marginal mines while being punitive on those which achieve high levels of profitability.

The emphasis on immediate revenue collection is a real one in the context of countries which are riddled with debt, and unable to provide a basic range of services, let alone create the material conditions for economic growth and development. In this context the predominant emphasis is to exploit the resources once a surplus can be generated, and make as large as possible front-end levies, which are independent of the degree of profitability of the project. Issues of revenue maximisation over time and conservation take a back seat to the central concern to generate immediate revenues. This approach is usually anti-conservationist even when price expectations indicate that the deposit should not be developed (that is, when the net price is rising faster than the rate of inflation). Front-end instruments such as royalties can affect the production schedule in terms of determining cut-off grades and the profitability of a marginal project. This issue is considered further later in this chapter.

The government may find out that its mineral endowment base is good, but it lacks the capital and the know-how to develop these resources and to market them and hence embark on a programme of attracting private investments. The usual methods that have been used to attract investment are the granting of special incentives such as tax holidays, lower tax rates, special allowances and early capital write-off provisions, and where the source of capital is sought from overseas, liberal rights to freely exchange currency and remit dividends. Because of the peculiarly capital-intensive nature of mining, the mere attraction of investment may not completely fit into the macroeconomic objectives of the country.

The predisposition of mining industries to an enclave status has been commented on by numerous observers, who have used it to either challenge the very relevance of mining in a developmental context, or to indicate the reasons for it, and hence positing suggestions as to what should be done to integrate the mining operations within the general economy (Girvan, 1976; Asante, 1979). The reasons for this enclave status are either to be found in the locational characteristics of mining (usually in remote hinterland areas), its capital intensity (hence higher levels of marginal product and wages), its economic and legal insulation (by special fiscal and legal arrangements), or perhaps the peculiar development of mining in a colonial context where, as it is posited, there is a direct link between the mine in the hinterland and the harbour from which the mineral products are shipped to the metropolis (Girvan 1970).

The fight against the economic dualism which exists between the enclave economy and the national economy has always been of major concern to governments of developing countries, and especially those in which a mining sector is present. A prevalent view is that in such countries mining industries which were principally controlled by the foreign investor should be made to play a more pivotal role in the transformation of the national economies. They should be integrated into the host economy in terms of the possibilities for meaningful linkages to other sectors.

The above concern has been reflected in various statutes of developing countries as was noted in the previous chapter, and is a policy preoccupation in its own right in the mineral sector. The state often combats this policy area by using the instrument of participation in the project, where it can speak as an equity holder, rather than purely as a regulator in its governmental role. It feels that, by so doing, it could give direct impetus to the immersion of the mineral sector in the total economy. Of course, its role as a regulator is pivotal in this growth area.

Conservation, though an old concept, found its way back to an area of prominence in the post 'Limits to Growth' period, where huge deficits of mineral resources were being projected to occur, because they were being depleted by an ever hungry, consuming world, which took little notice of their exhaustible (non-renewable) nature. Though the myth of impending scarcity has been seriously challenged, the world became aware of the need to 'use wisely' – that is to conserve, its natural resources (see Walrond, 1976).

In economic terms, one view is that conservation represents a state where the inter-temporal distribution of use rates of the resources is biased towards the future. The economic rationale for this preference of future over present consumption is related to the feeling that the net price of the resource in the ground is rising faster than the rate of inflation; hence the present value of the future income stream will be greater than that of the current income stream. Others see it as one part of a fully developed economic theory of resources (see Herfindahl and Kneese, 1974).

Fiscal instruments which are not neutral on different levels of an income stream can be either depletionary or conservationist. For instance, the levies which treat marginal projects poorly and bonanzas favourably are depletionary, while those that are of a highly progressive nature are conservationist in effect. This is an essential area of mineral policy and it has to be set against the other objectives in determining the fiscal package confronting investors in the sector.

The economic development concerns expressed in the mining laws of the under-developed countries, along with the prominence of such issues in the mineral development contracts which they have signed, as well as national economic policy pronouncements, reflect to a large extent their aspirations to attain sovereignty over their natural resources as conceived in the UN Declaration No. 1803 of 1962. The legal climate can assist in the realisation of that sovereignty, but it is the financial issues to be discussed in the next few chapters that will be crucial in terms of determining financial and economic feasibility without which sovereignty has little meaning. The mineral resource has to produce net surpluses of convertible currency, which can then be deployed in ways to meet the country's needs. These gains to economic welfare are essential to the entire issue of sovereignty.

Another aspect noted is that too little emphasis has been placed, even in the case of some of the more modern statutes, on the benefits of small scale mining to an under-developed country. Where capital is scarce, and expertise is limited, and if there is potential for mining development, it might be easier to put on stream a number of small- to medium-sized operations than to concentrate on putting together the 'big one'. The 'big one' ethic may pose insuperable difficulties for the state which invariably will have to face investments of the size sometimes of the entire national product, and the claims these will make on an already poor availability of skills, infrastructure, and services.

Legislation for the small sector needs further development. Each country needs to define what 'small' means in its particular circumstances, and programme the possibility of putting on stream a number of these over specified time frames. The legislation should be framed to meet the peculiarities of exploring for and developing deposits of that size. It may be inappropriate, for example, for a mine of 250 tonnes per day of 0.5 oz per tonne of, say gold, to have to put up proposals for training, infrastructure and economic linkages since the project will simply not be able to bear the costs. However, there should be provisions for the operation to run efficiently and use proper and safe techniques. Safeguards must also be made to prevent smuggling.

Small-scale mining, if effectively developed and policed, can net foreign earners as well as promote local entrepreneurship and regional development although it may not have the economies of scale and sophisticated technology that large-scale mining permits. A modified legal regime between the 'restrictiveness' of the three-tier systems and the 'looseness' of the claims systems described above, needs to be carved out. It might be necessary to give incentives for artisanal-style operations to be transformed into serious economic, though small, technologically sound ventures.

All the above objectives have to recognise that the principal economic justification for the various levels in the mineral sector is to capture the bulk of the resource rent generated by its exploitation, but without affecting operational efficiency or discouraging investment. The economic rent concept presupposes that all factors of production have been remuneratively (that is, optimally) employed, and therefore by inference, capital and entrepreneurship, which the investor represents, would be compensated. The distribution compared with project surplus between the investor and the government is determined partly by the fiscal regime, and this is taken up at length in chapter 4.

ROYALTY

Royalty is considered as a levy on the resource owner for the extraction of a non-renewable resource. The base for the levy is either on the output per unit, or on price, or on value of production or on a combination of either of these. It is to be noted also that some levies which are ostensibly made for access to the resource are calculated on the basis of profit (see Commonwealth Secretariat 1977b, for the different types of royalties and quasi-royalties which have been identified). Appendix I illustrates the types and rates of royalties levied by the respective states, while Table 3.1 summarises the ad valorem (or value) levies which are present in the countries. The point should be made that in as much as Table 3.1 indicates that there are no royalties in Zambia and Quebec, both of these countries levy a minerals tax which, though on a profit base, is in effect a royalty since it is a specific charge for the use of the resource.

Both Appendix 1 and Table 3.1 also show that even within a specific state, the rate of royalty varies for different mineral types, except in Quebec where no such distinction is made. This differential treatment of minerals reflects the difference in value per unit of output accorded to the minerals. For example, the high-volume, low-value industrial minerals typically have lower royalty rates applied, compared to the low-volume, high-value products such as precious stones and precious metals. This is a generalisation at best, since implicit in some of the royalty levies is the attempt by some states to collect a portion of any economic rent which may occur or it may simply represent the states' attempt to collect revenues regardless of such concerns. Notwithstanding these concerns, Table 3.1 does show that where ad valorem royalties are levied, the maximum level of the levy is about 5 per cent.

All royalties which are levied on a base which is independent of profitability can be viewed as an element of production cost. This therefore means that the level of the royalty would in fact affect the revenue-cost ratio of the venture. High royalties would reflect themselves in low revenue-cost ratios which, in turn, can lead to low

TABLE 3.1 Ad valorem royalty charges on selected minerals

	Sierra Leone	Tanzania	Zambia	Botswana	Papua New Guinea	Malaysia	Quebec (Canada)	Western Australia
Barytes	—	—		3	1¼	—		5
Coal, Lignite	—	—		5	1¼	—		d
Kaolin	—	2½	See Mineral Tax	3	1¼	—	See Mining Tax	5
Mica	—	5		3	1¼	—		5
Asbestos	—	—		3	1¼	—		5
Dimenite (rutile)	a	5		3	1¼	—		f
Bauxite	b	5		3	1¼	—		7½
Manganese	—	5		3	1¼	—		7½
Tin	—	5		3	1¼	c		2½
Columbium	4–50[b]	5		3	1¼	—		5

Lead	—	5	i	3	$1\frac{1}{4}$	—		5
Zinc	—	5	n	3	$1\frac{1}{4}$	—	i	5
Copper	—	—	A	3	$1\frac{1}{4}$	—	n	5
Platinum	5	5	p	5	$1\frac{1}{4}$			$2\frac{1}{2}$
			p					
Gold	5	$1\frac{1}{2}$	e	5	$1\frac{1}{4}$	5	A	5
Diamonds	5	15	n	10	$1\frac{1}{4}$	—	p	$7\frac{1}{2}$
			d				p	
Silver	—	$1\frac{1}{2}$	i	5	$1\frac{1}{4}$	—	e	$2\frac{1}{2}$
Tungsten	—	—	x	3	$1\frac{1}{4}$	—	n	5
						—	d	
Thorium	—	—	1(C)	5	$1\frac{1}{4}$	—	i	5
Zircon	—	—		3	$1\frac{1}{4}$	—	x	$2\frac{1}{2}$
							1(G)	

a Depending upon level of output (see Appendix 1(A))
b Depending upon profit
c Tin profits tax and export duty operate like royalty. See Appendix 1(G).
d See Appendix 1(H).

profitability and hence smaller surpluses available for distribution between the State and the investor.

Furthermore, when royalty is levied at the front end, and therefore 'production costs' notionally increased, the mining operator must increase the minimum grade of ore which he would otherwise have used to maintain the commercial viability of the deposit in terms of the loss to income caused by the marginal cost which the royalty represents. This can be extremely punitive on marginal deposits and can lead to their closure, or non-development, while high grading in other commercial deposits may reduce their longevity. This aspect of royalty is an essential point for governments to recognise when they contemplate making front-end levies.

If the general policy towards development of a mineral is conservationist, the use of royalty as an instrument is a two-edged sword; it allows for conservation (by postponing marginal developments and raising the cut-off grade, and hence reducing exploitable volume in commercial deposits) but it may also cause destruction of some deposits which may be exploitable only above a certain grade, because a lower average grade may translate itself into a higher average per unit cost of production, which may place the deposit within the range of entry of competing products. In that sense, the use of royalty as a conservationist or revenue maximisation instrument is sub-optimal.

Naturally though, front-end royalties provide a more assured source of income than levies based on profitability, since the determination of the level of the latter base is basically out of the government's control, and is only achievable depending on the state of all the variables entering the production process, for example, prices, weather, and mechanical efficiency. States have become wary of the integrated multinational mining enterprises shifting their cost centres and transferring prices of both inputs and outputs when possible, thus making the true profitability from the said operations very difficult to determine. Some states have further argued that the assured income stream represented by a front-end royalty is not a question of morality, but one of justifiable economics, where the royalty is viewed simply as the price of the natural resource as an input, like all other inputs for which payments are invariably made, in the production process. The fact that there is no foolproof way of assessing the value of the resource, however, leads one to wonder whether the latter approach might not be a sub-optimal one. It may seem more appropriate to view the surplus (economic rent discussed above) as the price of the resource since all other inputs, by definition, would have received their compensation. The argument becomes circular at this point, since the practical determination of a real compensation to the other inputs is far from being resolved. It is even further complicated by the difficulty of determining at what level the extraction of the economic rent will create a disincentive for operational efficiency.

Notwithstanding the difficulties in determining an acceptable level of royalty, some generalisations can be made if the government wants a mining project to be developed, and it must secure revenues by way of a front-end royalty levy. In the first place the levy should be made at a level which does not seriously disturb the financial viability of the project. A royalty on value of production is always better than one on unit production or in price, since it has in-built stability, and the government can benefit from any movements (especially if positive) in prices. Mineral prices are extremely erratic and, therefore, constitute the worst base on which a levy intending to secure a constant source of revenue can be based, except of couse, where a floor price for royalty determination is established. Naturally, if the floor price for royalty purposes is way above the real attainable price, the project finances can be disturbed.

INCOME-RELATED TAXES

The discussion in the last section identified taxes on income as a basic method used by governments to extract a return from economic activity in general, and in the case of mineral exploitation, as a tool for extracting a portion of the economic rent which accrues as a consequence of mining and disposal of the mineral products. At the outset, it should be recognised that the impositions under this head are established as some accounting concept of assessable income, which differs from one jurisdiction to the next. However, the basic concept is that a levy is made on income after deductions of some items considered as allowable costs of production, or other items which are intended to create an incentive, such as an allowance. The discussion on income-related taxes will therefore have to take cognisance of the level of the levies and the other accounting instruments which determine the base on which the imposition is made.

Table 3.2 indicates that the level of income tax in the various jurisdictions ranges from 22.5 per cent for certain categories of new mining projects in Tanzania to 55 per cent. A closer perusal of the table reveals that about 45 per cent is the usual level of basic company rate of tax used by most of the jurisdictions. These however, are not im-

TABLE 3.2 Comparative basic tax rates

Country	Basic company rate of tax %	Range of allowances Initial %	Annual %	Withholding tax on Dividends %	Interest %	Other taxes and remarks
Sierra Leone	45	5	(min) 5[a]	45	45	Surtax 15%; Diamond industry profits tax 27½%; Iron ore concession tax 5%; Tax holiday may be granted for up to 5 years.
Tanzania	22½[c]–50–55[b]	40	10	10	12½	Diamond levy of 5%; ad vaorem[f] sales tax of 24% on local sales of precious stones.[g]
Zambia	45	10	2[d]–30	20	20	Mineral tax (on assessable income): Cu 51%; Aw, Bi, Se, Co, Ag, Cd 10%; Pb 20%; Zn 20%; Amethyst/Beryl 15%.
Botswana	35	Nil	—[e]	15	15	
Papua New Guinea	33⅓–45[g]	25	5–20[g]	15	45	Additional Profit Tax (APT) of (70−n) ΣNCR[g]

Malaysia	40	20	80	15	0
Quebec (Canada)	12[h] 36	—[i]	30[i]	—	—
Western Australia	45	47½		30[j]	10

a There is an annual allowance on 'Qualifying mining expenditures' which is calculated as the greater of 1/20th of the basic period year's output divided by the basic year's output and future potential output.
b Non-resident company permanently established.
c Income derived from precious stones mining for first four years only. The other rates apply from the fifth year. For specified minerals (Cu, Au, Sn, W, $CaCO_3$, $MgCo_3$, mica) the rate for the first 4 years is 45%, thereafter 50%.
d Capital expenditure incurred by 'new mines' is allowed in full in the year incurred. For old mines, plant and machinery 30%, industrial buildings 5%.
e Residual capital expenditure including capital expenditure incurred during tax year, divided by expected life of the mine or thirty (whichever is less).
f If gross production is less than 10000 shillings
g NCF: Net Cash Flow; See Appendix 1(E).
h 36% (that is, 46% basic company tax minus 10% rebate) represents the Federal corporate tax and 12% represents the provisional (Quebec) corporate tax.
i On development expenses, see Appendix 1(G) for the range and definition of allowances and taxes.
j Reduced to 15% where double taxation treatise concluded.

Development tax – 5% taxable profits (see Appendix F for incentives).
Quebec mining tax graduated from 0 to 30% (see Appendix 1(G)).

mediately comparable because of the differences in the computation of the taxable base, and the fact that some jurisdictions which may have low basic company tax rates, also use other income-related levies such as a development tax in Malaysia, a mining tax in Quebec, a mineral tax in Zambia, a diamond profits tax in Sierra Leone, or an additional profits tax as in Papua New Guinea. Some basic deductions recognised in all jurisdictions are operating cost, interest, depreciation and royalty, where applicalbe, in the computation of assessable income. The income tax provisions of Quebec are the most complex since different bases are established for the computation of the federal tax, the provincial tax and the Quebec mining tax (see Appendix 1 (G)).

The computation of the tax base and the level of the tax in the various jurisdictions can normally be seen to reflect the state's perception of the prospects for minerals in the country. For instance, Tanzania levies an extremely low level of tax at $22\frac{1}{2}$ per cent for diamond operations in the first four years, while imposing a 45 per cent tax on all other mining activities during that period. This reflects the obvious attempt of the Tanzanian Government to promote mining. It goes even further by placing a high level sales tax of 24 per cent on diamonds sold locally, thus reflecting the State's desire to have the diamonds exported to gain valuable foreign exchange. Diamonds are obviously well entrenched in Sierra Leone, as the industry has to pay a surtax of 15 per cent plus a $27\frac{1}{2}$ per cent of diamond industry profits tax, apart from the basic 45 per cent company tax. Zambia is very dependent on income from copper operations, which represent the major source of foreign exchange earnings and government revenue. The government therefore levies a 51 per cent tax on assessable income from copper operations, as opposed to mineral tax of 10 per cent for selenium, cobalt, cadmium and silver operations, which are to be encouraged. These, of course, being additional to the normal corporate income tax of 45 per cent.

Though most jurisdictions attempt to levy the same level of company tax for all sectors of the economy, the computation of the mining company tax base is quite different. This perhaps reflects the fact that all the jurisdictions recognise that mining may be set apart from other sectors, principally from the standpoint of the exhaustible nature of the asset, the question of sovereignty as it is reflected in resource ownership, the relative potential importance to the economy of single large mining venture, the geological and other risks involved in searching for, developing and exploiting of a mineral deposit, and the fact that large mining investments may not be realisable if the activity had to be prematurely stopped for any reason.

The foregoing factors are usually accommodated in the treatment of mining expenditure and the allowances granted to promote and maintain mining. Expenditure incurred on exploration, development and continuing mining operations are basically the major recognisable capital expenditures to be amortised by the company. However, various jurisdictions treat them differently. For example, Tanzania allows all capital expenditure including exploration to be deducted on a schedule of 40:10:10:10:10:10:10:, while Zambia treats all capital expenditure on new mines as deductible in full in the year incurred. For old mines, Zambia distinguishes between expenditure on plant and machinery (30 per cent on a declining balance), industrial buildings (10 per cent initial and 5 per cent on annual) and low-cost housing (10 per cent initial and 10 per cent annual). Botswana, on the other hand, permits capital write-off over the life of the mine, which at any rate should not exceed 30 years. In the case of Papua New Guinea (see Appendix 1(E)), apart from the distinction between scales of mining operations, and the computation of annual allowances based on the ratio of the residual expenditure to remaining life of mine, it is to be noted that exploration expenses are treated differently from development expenses, and in fact are allowed at a greater rate (maximum of 20 per cent exploration expenses compared to maximum of 10 per cent for other capital expenditure). Quebec also makes a distinction between exploration and other pre-production costs, while also distinguishing between pre-production and post-production expenditure (see Appendix 1 (G)). Table 3.3 summarises the capital allowance structure of selected developing countries while Appendix 4 outlines the key depreciation methods used.

An analysis of the above permits some general observations, which reflect the reasons for the choice of certain instruments by the various states. Though basic company rates were observed to be similar across all sectors of the economy, states have used the vehicle of tax depreciation to accord either special treatment to the mining industry or simply to treat with its peculiar nature. There was evidence of quick write-off provisions of capital expenditure, which could only be designed to counter the high risks associated with mining. The opportunity for a quick recoupment of funds is a powerful incentive for investment in exploration and development (see Kumar and Walrond, 1983, for a deeper treatment of this issue). Generous depletion allowances as in Quebec are yet another form of creating an incentive for mining. Some states (Papua New Guinea and Malaysia) have used the tax rate and the computation of the tax base as a means of giving preference to local investors, while others, like Botswana have attempt-

TABLE 3.3 *Capital allowance structure for mining*

Country	Capital allowance structure
Sierra Leone	Greater of 5% as (output/reserves)
Tanzania	Deducted on a schedule of 40:10:10:10:10:10:10 including write-off of exploration. For 'specified minerals' deduction is allowed as expenditure is incurred.

Zambia	For old mines:
	  Plant & machinery: 30% on a declining balance
	  Industrial building: 10% 10% initial
	      5% annual
	  Low-cost housing: 10% initial
	      10% annual
	For new mines:
	  Capital expenditure deductible in full in year incurred

Botswana	Written off over life of mine, which should not exceed 30 years
Papua New Guinea	Initial allowance: maximum of 25% Annual allowance: Residual Expenditure (RE) or $\frac{RE}{\text{Remaining life of mine}}$ or $\frac{RE}{10}$ whichever is less. Therefore maximum of 10% for capital expenditure. Annual allowance (for previous exploration) $= \frac{\text{Residual Exploration Expenditure}}{\text{Remaining life of mine}}$ or $\frac{REE}{5}$ Therefore, maximum of 20% for exploration expenditure
Malaysia	For the period 1978–86, all industries are given an 80% accelerated allowance on plant machinery apart from the 20% initial allowance already in place (that is, effectively immediate write-off)

ed to tie write-off provisions to the life of the mine, much akin to the standard procedure of allowing depreciation over the expected life of a piece of equipment. Sierra Leone has provisions for a tax holiday for a prescribed period when the state consideres it necessary for development of a project.

The important point of whether or not to give any form of tax incentives should be approached with caution, since they could distort the equity in the tax system, pose administrative difficulties and provide savings marginal to the decision to invest and an unnecessary source of revenue erosion.

Immediate expensing tax provisions for capital, coupled with provisions for carrying forward losses, benefit the investor since the government's take via taxes will only come into force when the enterprise makes a profit. The tax holiday incentive on the other hand, does not only reduce the government flows over the life of the project, but invariably involves transfers to the investor at an early stage of the project, when net discounted present values are higher. From the government's viewpoint, tax holidays do not allow the state from capturing any profits which may arise from the mining operations. As such, the state's interest could be best served by granting accelerated write-off, but allowing itself the opportunity of collecting some taxes from the early years of the project. In the case of mining, where other economic benefits such as employment creation will be limited, the state will get very little from the mining operations if it gives a tax holiday. What is worse is, that barring tax-sparing provisions in a double taxation agreement between states, the intent of the tax holiday is lost, since the investor's home country will simply tax the extra income leaving the investor no better off. In fact, the investor's home could be the real beneficiary to a tax holiday in these circumstances. Recall that, under tax-sparing provisions, the home country government of the transnational allows it to regard foreign-source income as if host country income taxes had already been paid, even when the host country grants a tax holiday, as, for example, in Holland. In sum, an accelerated capital allowance scheme as well as a tax holiday therefore gives immense benefit to the investor in terms of front-end liquidity, thus enabling the project to service loans and repay debts and sometimes declare dividends before income tax itself becomes due. This will affect the ability of the project to attract financial capital. They can, therefore, reduce the risk of investment and lower the supply price of investment. Accordingly, it is an important factor in pushing ahead the marginal project if it is desired in itself. For the more profitable project, however, faster write-offs can favour the government, *ceteris paribus*, where there is a profit-based rent resource tax, because they can help to speed up the payback period as well as enable the project to achieve a faster target rate of return. (The Papua New Guinea case in the financial model in chapter 4 will elucidate this.) Of course, as discussed above, this will depend on how effective and efficient the income tax machinery is in monitoring costs and revenues and collecting the tax. However, when there is no rent resource scheme, and if the project is a profitable one, the government could end up giving a concession which might not be necessary at all. Although accelerated capital allowances do not affect the total amount of government revenue over the life of the project, only the timing, this is only in *nominal* terms. Given that money received early is more valuable than later, there is cost to government in net present value terms in providing the investor the accelerated depreciation concession if the project is profitable (see also chapter 4 for a quantitative verification).

It makes sense, therefore, for government to make accelerated depreciation a flexible tool to bargain against other elements in the fiscal package. It is an important card in the government's hand. If there is economic potential for further investment in the mining project, another safeguard would be to link the accelerated depreciation concession to reinvestment guarantees. A stipulated minimum percentage of the write-offs granted for equity capital could be reinvested in the mining project or other economically viable projects in the country. The stipulation might also be that some part of the write-off capital be reinvested in new exploration or in the project itself. While the size and nature of the project and the expected economic rent that will be generated will determine the capital allowance scheme, it is important that any accelerated depreciation scheme should be a *conditional* one, and not an open-ended concession. Another point to note is that in giving accelerated depreciation, government has to ensure that there are other fiscal elements, such as a royalty, to give it a steady flow of income not entirely dependent on profitability. Otherwise it could end up receiving nothing in the early years of a project that is earning profits but not paying any tax because the investor has yet to achieve the target rate of return.

Apart from the corporate tax mentioned above, a significant income-related tax which can seriously affect the real rate of return to the investor is the withholding tax. As will be illustrated in chapter 4 (last section), the return to the investor is always smaller than the return to the project when repatriated dividends are subject to a withholding tax. The range of withholding taxes in the various jurisdictions is from 15 per cent to 45 per cent, with most states opting for the 15 per cent provision (see Table 3.2).

The intention of the dividend withholding tax is to collect tax on income received by shareholders abroad from income produced in the host country. At the same time, it acts as a 'stick' to reinvest earnings; and in a sense an alternative to reinvest incentives. If reinvestment does not take place, the withholding tax gives the government revenue from divestment.

Though it does not have an income base, a withholding tax on interest is a standard instrument used by most states where offshore financing is involved. To the extent that interest on loans is a first charge on income before tax, the parent of a subsidiary is more inclined to receive its return by way of interest, especially in circumstances where a dividend withholding tax is in place, and the corporate tax rate is high. This is so because the interest income is then tax free, and the return to the lender greater than if he had received his return by way of part-tax dividends which may also be taxed.

To illustrate the incentive to disguise equity as debt in circumstances of a dividend withholding tax of 15 per cent and an income tax of 40 per cent, it is assumed that a US$10 million investment produces US$1.5 million before tax, that is, a 15 per cent rate of return. The position of two alternative cases of 100 per cent equity financing, and 100 per cent debt financing, is as follows:

	US$10 million as equity from parent company	US$10 million as loans with 10% interest (debt)
	$	$
Net income before tax	1.50 m	1.50 m
Less interest of	0.00 m	1.00 m
Income tax at 40%	0.60 m	0.20 m
Dividends	0.90 m	0.30 m
Dividend withholding tax at 15%	0.14 m	0.05 m
Net take to government (income tax & dividend withholding tax)	0.74 m	0.25 m (+0.15 interest withholding tax)

It can be seen that the tax that will be collected by government will be higher if the $10 million investment had been labelled as equity instead of debt. Even if there is an *interest* withholding tax of 15 per cent the extra revenue will only be $0.15 million in the above example, giving a total of $0.40 million which is still lower than a 100 per cent equity case. There are many ways of tackling this problem. One is *not* to allow interest as a deduction expense, particularly for debt held by affiliates but this could be a burden where debt servicing is essential for the viability of the project. A better way would be to disallow interest deductions on long-term loans that bring the debt-equity ratio above a prescribed figure, for example 7:3 or 3:1.

States have to be alert to this form of abuse and therefore put in place mechanisms to prevent, or at least restrict, this kind of transfer of income. Policy with respect to the gearing of the project, and the rules governing associated companies have to be well thought out. The various jurisdictions levied interest withholding taxes of between 10 and 45 per cent, with Malaysia not having any such provision.

Income taxes and royalties apart, various states have used other levies to increase government revenues or to give effect to some aspect of government policy. The diamond industry profits tax, the surtax of Sierra Leone, the development tax of Malaysia and the diamond levy of Tanzania have all been mentioned before. The local sales tax of Tanzania was designed to induce exports of diamonds while the mineral rights tax of Botswana was designed to appropriate for the state all non-beneficially occupied mineral lands. All non-income related levies are akin to royalties, and their effects have been discussed in the previous section.

RESOURCE RENT TAX

The discussion thus far has centered on the collection capability of the State via the mechanism of income-related taxes and royalties. It further attempted to address the question of equity of treatment across various sectors. However, the question of the efficiency of the instruments to achieve government policy to maximise revenues from mineral exploitation has to be tackled. In this context, the measure of efficiency will be the ability of the tax to permit marginal operations to be viable, while withdrawing the maximum economic rent from a bonanza.

A resource rent tax (RRT) above the normal profit that investors can earn elsewhere has been justified because it is the nature of the resource-endowment and location that produces the surplus. According to economic theory, the entire surplus over and above the normal profit necessary to attract the investment (that is, the supply price of investment) should be taken away as rent by the host country, except the premium necessary for risk aversion. The argument is, that, since the incentive by the investor to do business is not affected, that is, it leaves normal profit untouched, the tax is neutral in its effects, and not distortionary.

While the theory of economic rent is easy to understand and is less controversial, the practical application of the tax is beset with a whole host of complexities that range from how to measure economic rent to difficulties of monitoring and collecting the tax.

The resource rent tax, although it has gained renewed popularity since the 1970s, is not a new form of tax. Countries such as Colombia and Argentina have used the tax since before the Second World War, although developments in the 1960s have changed its character.

The Valco Fund, set up under the 1962 Agreement to establish the Valco Aluminium Company Limited in Ghana, is a form of economic rent transfer whereby 50 per cent of the excess profits made after deducting various prescribed items, including a return on equity, is given to a Fund set for cultural, educational and scientific purposes.

The 1970s saw new forms of resource rent schemes variously labelled as Additional Profits Tax (APT) in Papua New Guinea and Tanzania, Corporate Income Tax on windfall profits in hard minerals in Indonesia, Petroleum Revenue Tax (PRT) in the United Kingdom, Additional Mining Royalty in Guyana, and the 'price cap' in Angola.

The mode of operation varies from regime to regime. A popular version that is exemplified in some agreements of Papua New Guinea, Tanzania and Guyana is basically a year to year *cash* flow concept, and the RRT is imposed on the net cash flow of the mining company. From the total revenue of the project each year, total payments that include exploration expenditure, development and other capital costs, operating costs, royalties and income taxes are deducted. (In its pure form, there are no income taxes.)

The net cash flows are carried from year to year, increasing at what is normally referred to as the *threshold rate of return* or accumulation rate, which is agreed by the host government and investor. In the early years, the net cash flows remain negative since the entire capital invested is deducted from the gross revenue. As profits are added to the receipts every year, the negative cash flows diminish, and as soon as they become positive the resource rent tax of say 50 per cent is imposed. Accumulation of the net cash receipts by the threshold rate of return only proceeds as long as the net cash receipts are negative. This allows the company to recoup its original capital as well as earn a normal rate of return (usually averaging 15 to 20 per cent nominal) before paying the resource rent tax. If there is no income tax, the RRT gives the multinational a tax holiday for the pay-back period. Unlike a predefined period of tax holiday usually granted by less developed countries to promote investment, the tax-holiday of the RRT system is decided *ex post* by the net cash flow. (The tables in Appendix 3 illuminate selected computations including the effect of the **RRT** in the case of Papua New Guinea. Appendix 5 contains an example of a **RRT** calculation.)

There are several practical problems in implementing the tax, and international experience of the **RRT** suggests that the amount collected by this source has been generally low. Deciding on the 'threshold' rate of return, more often than not has been arbitary, and rates that have been chosen range from 15 per cent to 20 per cent for hardrock minerals. The accumulation rate is a matter for negotiation, and generally incorporates a risk premium.

Once the rate has been decided, problems could arise in monitoring the costs and profits of the company. Profits shown on books may be grossly understated, and costs inflated. For the transnational, the mechanism of transfer pricing provides an avenue for tax manipulation. For the collection to be effective, the tax collecting machinery of the less developed countries must be relatively sophisticated and well-versed in detecting transfer pricing. Indeed, industrialised countries like Britain and Norway with a well-developed income tax structure are said to be facing difficulty in administering their respective RRTs. The more sophisticated the structure, the more difficult it becomes to administer practically. Herein lies the dilemma. In the accounting sense, there is still controversy as to whether the invested capital should be defined in original cost terms, replacement cost terms or book value. What happens to leased capital – should it be defined as part upon which the rate of return is calculated? Should interest be an allowable deduction? Does the debt-equity structure of the company matter? Should the tax be creditable in the home country?

These issues are by no means conclusive but indicative of the practical difficulties of implementing a resource rent tax. These difficulties are certainly greater than in the case of royalties or an output-based tax. As Gillis and Wells (1980) comment, the simpler the structure of the tax, the weaker will be the safeguards that might limit tax avoidance, whether through transfer pricing devices, inflation of the capital base upon which the RRT is to be computed, or any of the many other techniques known to any competent manager of a multinational enterprise. The more theoretically cogent it is, the more difficult it becomes to administer. It therefore makes sense for the tax collecting machinery of developing countries to be staffed with well-trained, well-paid and highly motivated people backed by technical know-how of the mining industry to ensure that the appropriate tax dues are collected.

OTHER FISCAL DUES

It is common for the less developed countries to grant full or partial customs duty exemptions on all imports of capital goods, raw materials, and in some cases all supplies, including food and cars for expatriate officials for mining projects. While this concession can be supported in cases where the goods are not available locally but necessary for the production process, it is less justified if duty free privileges are given on the import of raw materials and supplies available domestically. There are examples of mining companies importing basic items like soap and soft drinks duty free even when local supplies are amply available. Revenue evasion of this kind should not be encouraged as local sourcing can have benefits in terms of foreign exchange and regional development.

Other levies that have been imposed include land and property tax, sales tax, pay roll taxes and stamp duties. Of direct relevance to mining is land rent. This varies from country to country. Land rent should be progressive over time, that is, its rate per hectare (or acre) increasing from year to year with the aim of discouraging companies to hold on to land that they are unable or unwilling to exploit. This is especially desirable if there are no relinquishment provisions specified in the agreement. Another important feature is that the rates should be index-linked to maintain their real values.

4 Analysis of Impact of Regimes

THE MODEL

The model that has been chosen is taken to resemble that of a gold mining project, but it has been simplified and structured in such a way so as to have general applicability. In casting the model it is recognised that the treatment of case studies can generally be restrictive in their applicability to other forms of mining, since by their nature each mineral project can be considered unique. However, to assess the impact of various fiscal provisions as they are enshrined in the respective statutes of the countries, the assumptions that have been made seek to hide the identity of the mineral, and hence render the results generally applicable.

(a) The first assumption is that the relationship of revenues to production costs (both capital and operating) is one of the key elements in the decision with respect to financial viability and the quantum of the project surplus that would be divided between the government and the investor. Regardless of whether the mineral is gold, copper, tin or coal, the relationship exhibited between the ratios of revenue to cost over time is the fundamental one.

(b) The scale of the project has significant impact on the subsequent financial flows. The financing requirements of large projects may involve restrictions which would further complicate the actual analysis of the project. However, regardless of whether the project deals with revenue of a hundred million or a thousand millions dollars, the fundamental dynamic relationship of revenues to costs is still an overriding criterion.

(c) Modelling all cost elements and/or price changes can result in an unmanageable number of permutations, resulting in an infinite number of analyses. Hence, this model should be read in the context that the high-price scenario can be construed as a low-cost one, and conversely, the low-price scenario, which implies that the ratio of costs to revenues is high, can be regarded as a high-cost scenario.

Analysis of Impact of Regimes 57

(d) Costs and revenues are assumed in the financial sense, not taking into account their 'shadow prices'. Furthermore, the project will derive other benefits such as employment, skill improvement, regional growth, income tax from employees and a host of other linkage and multiplier effects. On the cost side there could be externalities in terms of pollution, exchange rate effects and other macro distortions. To incorporate all these will require a social accounting or cost-benefit form of appraisal. This form of appraisal, besides its problems of quantification, is still quite controversial. Furthermore, getting down to the essentials of linkage and multiplier effects could distract from the objectives of the analysis, which is mainly intended to determine commercial viability.

ASSUMPTIONS OF THE MODEL AND DEFINITIONS

The above review of the mining developments in the countries being analysed as well as others reveals that there are basically four kinds of arrangements being contemplated in the developing countries which take their statutory provisions as a base. (It is recognised that there are other forms of contractual arrangements such as production sharing, and turn-key arrangements, and the model could be easily adapted to take into account other key elements.) The forms chosen for financial modelling are:

CASE I: 100 per cent equity ownership by the foreign investor with the host government receiving its take in the form of royalty, income-related taxes, withholding taxes on interests and dividends and other fees. The issue of leverage is ignored, and the investor is assumed to provide all funds. (Note that the investor could also be a private investor domestically based except that he may not be liable to dividend withholding tax.)

CASE II: The host government contributes 51 per cent of the equity of the venture with the investor taking 49 per cent. (Majority participation on which this is based has been noted to be reflected in government taking 51 per cent.) In addition to various fiscal payments, the host government will receive its share of the dividends from the project. (Note again that a domestic private investor could be a party in the venture.)

CASE III: The government has 51 per cent of the equity of the venture, of which 20 per cent is 'free' in the sense that it is paid for by the investor, and the remaining 31 per cent is fully paid for. (The concept of 'free equity' will be taken up later.) The foreign investor has 49 per cent. Furthermore, a 70:30 debt-equity ratio is assumed for all capital costs, that is, accumulated exploration costs and development costs. Interest on the loan of the project is 4 per cent real or 11 per cent nominal, and is capitalised in the year of production, that is, 1986. The loan which is repaid over six years (1986–91) in equal instalments is assumed to be raised overseas, and wherever applicable, the interest paid is subject to interest withholding tax.

CASE IV: The host government has 51 per cent of the equity of which 20 per cent is 'free', while the remaining 31 per cent is paid for by the investor and subsequently repaid out of 50 per cent of government dividends at an interest rate of 4 per cent real or 11 per cent nominal. The leverage of the project is the same as for Case III.

The other common assumptions of the model are:

(i) *Exploration Costs* are accumulated at the start of the development phase and are assumed to be US$5 million real. There is no new exploration activity in the development and production phases of the project. This is done to simplify the analysis.

(ii) Total *Development Costs* are assumed to be US$32.5 million in real terms, and are incurred over a three-year period, 1983 to 1985 inclusive.

(iii) *Production* begins in 1986 and lasts for 15 years, that is, until the year 2000. A medium production scenario is assumed, and it comprises 150 000 tons of ore at 0.4 oz per ton of recoverable gold per year, that is, 500 tons per day. Based on this, the recoverable gold is 60 000 ounces per annum. Total recoverable gold reserves in ounces are therefore 900 000. All output is assumed to be sold.

(iv) *Operating Costs*. These include both mining and milling costs and are assumed at US$8.80 per ton of ore mined and milled. Working capital and replacement capital are also subsumed under operating costs.

(v) *Prices* are assumed at US$300 per ounce, US$450 per ounce and US$600 per ounce.

(vi) *Inflation* is assumed at 7 per cent per annum throughout the planning period of both costs and prices. (It is useful to note that, if taxation and debt financing are ignored, uniform inflation in costs and revenue will produce a current money cash flow which itself increases by the inflation rate. In this study, taxation significantly influences the effect of inflation on discounted cash flow investment criteria. This influence results because tax allowances are usually based on actual historical costs as they are incurred. Due to debt financing, an evaluation based on constant money which ignores inflation will produce a rate of return on equity that is less than what is actually realised. The extent of debt financing may offset those produced by taxation, and the extent of offset will depend on both the borrowing terms and the mining taxation rules (see Mackenzie, 1983).)

(vii) *Discount Rate* is 5 per cent real, or 12 per cent on nominal values.

(viii) All foreign investor *dividends* are assumed at be remitted, and therefore attract dividend withholding tax wherever applicable. (Realistically, some of the profits may be reinvested; some kept as retained earnings.)

(ix) There are no lags in receipts and expenditure. This means that taxes, royalties, interest, dividends and withholding taxes are paid in the year of assessment to simplify computations. (Note that lags in payments of fiscal dues can increase the investor-discounted real net flows and internal rate of return, and correspondingly reduce government-discounted net take.)

(x) A common currency is used, that is, US dollars.

(xi) For simplicity, the project has no salvage value, that is, the value of the capital asset at the end of the planning period.

(xii) The ultimate objective is *profit maximisation*

Other important points to note are:

(a) Host government is being treated as a composite entity, both as a participant in the project as well as the tax collector. In reality, the participant could be a government-owned company.

(b) All income-related taxes in this study include corporation tax, surtax, development tax, mineral tax, federal tax and state tax where applicable.

(c) The respective internal rates of returns (IRRs) are computed from year-to-year flows throughout the 18-year planning period (1983 to 2000).

(d) The project equity is essentially that portion of capital costs (exploration and development) not financed by loan.
(e) None of the key parameters is subject to a probability form of analysis. This is because the probabilities will vary across the countries concerned due to different economic, geological and political risks. In this analysis, instead of the expected value approach, risk is included indirectly in the discount rate for calculating the net present value (NPV). The discount rate used is 5 per cent real or 12 per cent nominal, 7 per cent being the inflation rate. Thus the measure of attractiveness, in this instance the project NPV, contains an indirect or implicit factor which reflects the project's risk. No attempt is therefore made to express the risk of each proposal or portfolio of proposals. It is useful to point out that the risk assessment in this case is subjective, and to some extent arbitrary, and this is usually the case for different mining companies. Technically speaking, by adding the premium to the discount rate, the risk premium will itself grow with time. This assumes then that the risk does, in fact, increase over time at exactly the same rate as the discount factors diminish (see Levy and Sarnat, 1982, chapters 9 to 13 for an excellent discussion).

The computer results of the four scenarios permit investigation of the following:

(A) It allows the precise understanding of the concept of government participation and its implications for the government's share of the surplus of the project and the returns of the investor.
(B) It illuminates the real financial costs to the company of financing part of government equity while at the same time it allows one to establish the cost to the government of its own participation, whether by upfront contributions or from its future dividend stream. The impact of other than equity financing on the project flows can be gauged.
(C) The different price scenarios permit one to evaluate the impact of the various revenue/cost relationships to the cash flows and the ultimate return to the government and investor.
(D) The dynamics of the the various fiscal elements such as royalties, income-related taxes and depreciation schedules under the four scenarios and the various cost/revenue conditions can be understood.
(E) The examination of the impact of the total fiscal package on project of different degrees of profitability is permitted. Implicitly, the neutrality of the fiscal package in investment decision-making is being assessed.

It has to be emphasised that no attempt is being made to rank the legislative fiscal elements of the various countries in terms of severity to the investor. This is because mining agreements between host government and foreign investor might not necessarily include all the elements on the respective legislative schemes, and will vary from case to case depending on circumstances prevailing during negotiations. Purely judging the severity of the regimes can therefore be misleading. The issue is further taken up in chapter 5.

At this juncture it may be pertinent to mention that the financial analysis of mining projects has been moving towards new frontiers through the use of computers. The model above is a simple one, choosing data and parameters sufficient to illustrate the technique of computer-assisted financial analysis, and the policy implications based arising from them. There is now a greater emphasis on the use of computers throughout the entire evaluation cycle, from the initial geological model to the detailed feasibility analysis and distribution of economic rent. Appendix 6 contains an explanation of some basic financial concepts.

Key Definitions Identifying Actual Elements of the Computer Model

1. Gross Income Before Taxation (GIBT) = Gross Revenus = Output × Price
2. Net Income Before Taxation (NIBT) = Gross Revenue − Capital Allowance − Royalty − Operating Costs − Interest − Carried forward losses from previous year
3. Income Tax (IT) = NIBT × Income Tax Rate (%)
4. Net Assessable Revenue (NAR) for APT calculations = Gross Revenue − Exploration Cost − Development Cost − Royalty − Income-Related Taxes − Operating Costs
5. Accumulated Net Assessable Revenue (ANAR) = NAR + (last year's NAR times accumulation rate to trigger APT) and if positive equals NAR only
6. Additional Profits Tax (APT) = APT rate × NAR (only if positive)
7. Venture Dividends (VD) = Gross Revenue − Exploration Costs − Development Costs − Operating Costs − Royalty − all Income-Related Taxes − Loan Repayment + Loan Interest

8. Investor Dividends (ID) = VD × Investor Equity Share
9. Government Dividends (GD) = VD × Government Equity Share
10. Total Surplus (TS) = Gross Revenue − Exploration Costs − Development Costs − Operating Costs − Interest (net of Interest Withholding Tax) + Loan Repayments

NOTE: Total surplus is essentially the amount less all costs (except fiscal dues) that is shared between the two parties. The investor's share will be basically the dividends, whereas that of the government represents all fiscal dues as well as dividends if it is participating in the project (as in Cases II, III and IV).

In the summary tables the respective shares are listed as real government share of surplus for the government's portion, and real investor share of the surplus for the investor's. Detailed definitions of the two are as follows:

11. *Investor Real Net Cash Flow* (IRNCF)
    Cases I and II
    IRNCF = Investor Equity Share (%) × (Gross Revenue − Exploration Costs − Development Costs − Operating Costs − All Income-Related Taxes − Royalty) − Dividend Withholding Tax

    Cases III and IV
    IRNCF = Investor Equity Share (%) × (Gross Revenue − Operating Costs − Loan Interest − Royalty − All Income-Related Taxes − Loan Repayment − Project Equity) − Dividend Withholding Tax − 'Free Equity' to Government − Government Equity Carried by Investor + Government carried Interest − Equity Repayments to Investor including Interest

12. *Government Real Net Cash Flow* (GRNCF)
    Case I
    GRNCF = All Income-Related Taxes + Royalties
    Case II
    GRNCF = Government Equity Share (%) × (Gross Revenue − Operating Costs − Exploration Costs − Development Costs − All Income-Related Taxes − Royalties) + Dividend Withholding Tax + All Income-Related Taxes + Royalties

Cases III and IV
GRNCF = Government Equity Share (%) × (Gross Revenue − Operating Costs − Loan Interest − Royalty − Loan Repayment − All Income-Related Taxes − Project Equity) + All Income-Related Taxes + Royalties − Government carried Interest − Equity Repayments to Investor including Interest

13. *Real Government Equity Flows* (RGEF)
Case II
RGEF = Government Equity Shares (%) × (Gross Revenue − Development Costs − Exploration Costs − Opérating Costs − All Income-Related Taxes − Royalty)
Case IV
RGEF = Government Carried Equity (%) × (Gross Revenue − Project Equity − Loan Repayments − Operating Costs − All Income-Related Taxes − Royalty − Interest)

14. Project Surplus = Gross Revenue − Project Equity − Operating Costs − Loan Repayments − Interest − All Income-Related Taxes − Royalties

15. Government Take (%) = $\dfrac{\text{Government Real NCF}}{\text{Total Surplus}} \times 100$

THE RESULTS

Case I − Results of Model with 100% Fully Paid Equity by Foreign Investors

Table 4.1 summarises the results of the regimes for the three price assumptions. The most striking feature is that despite the identical cost-revenue assumptions for each country, the returns to the project, government and investor, vary for all corresponding prices. For the US$300 per oz case, Government's share of the real total surplus varies from a low of 37 per cent in the case of Quebec (Canada) to 100 per cent for Sierra Leone assuming the project is allowed to complete its 15-year production phase. Likewise the internal rates of return for the investor on his real net cash flows also vary, 9.7 per cent for Quebec and nil for

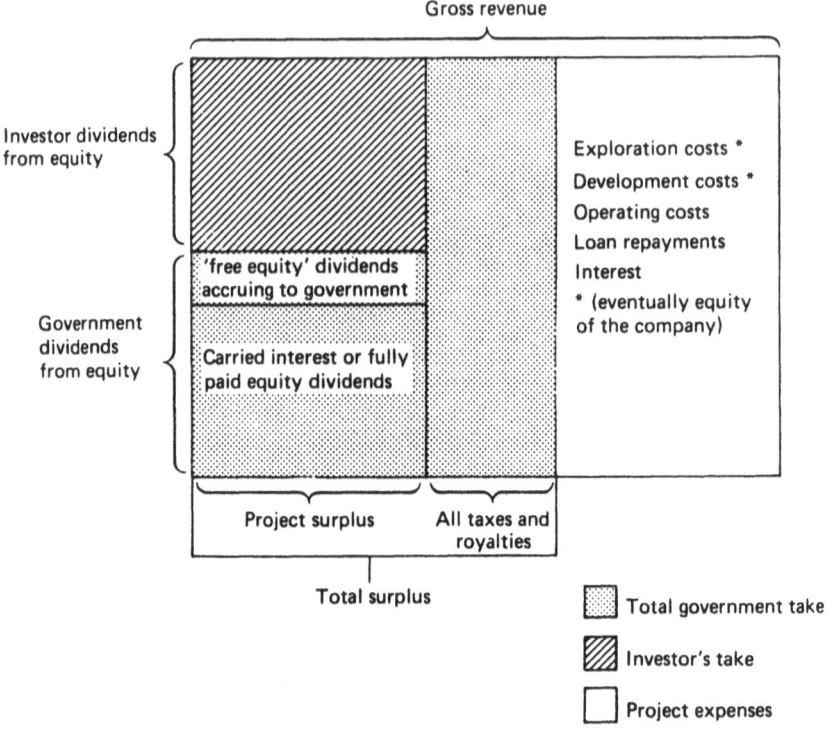

FIGURE 4.1 *Project flows and splits between government and investor*

Sierra Leone. The others fall in between the ranges. *The varying returns are purely due to the elements of the fiscal package as the IRR of the project confirms.* Whether the project would be undertaken is assumed to depend on the rate of return expectations of the investor. If it is 10 per cent real, none of the regimes is attractive. At 5 per cent real, Tanzania, Papua New Guinea and Quebec become possible.

An increase in the price to US$450 and US$600 results in the following:

(i) An increase in the returns in absolute amount to all parties in all countries. This is because for the same level of capital and operating expenditures, more is earned in revenue through the higher prices.

TABLE 4.1 Results of model with 100% fully paid equity by foreign investor

			Government				Investor		
	Real total surplus (RTS) (US$ (m))	Real project IRR (%)	Share of RTS (US$ (m))	Share of RTS (%)	Share of discounted (5%) RTS (US$ (m))	Real IRR on fully paid equity (%)	Share of RTS (US$ (m))	Share of discounted (5%) RTS (US$ (m))	IRR on real net cash flows
Gold Price: US$ 300/oz									
Sierra Leone	55	2.5	66*	100.0	41	—	−11	−18	0
Tanzania	55	6.7	39	70.9	22	—	16	0.2	5.1
Zambia	55	3.7	53	96.4	31	—	2	−9	0.7
Botswana	55	6.0	42	76.0	25	—	13	−3	3.8
Papua New Guinea	55	8.0	35	63.6	21	—	20	2	5.7
Perak (Malaysia)	55	6.0	44	80.0	25	—	11	−3	3.6
Quebec (Canada)	55	9.7	20	37.1	11	Quebec govt share of RTS (%) 11.7	35	12	9.7
Western Australia	55	—	—	—	—	—	—	—	—
Gold Price: US$ 450/oz									
Sierra Leone	190	15.1	167	87.9	102	—	23	3	6.1
Tanzania	190	20.8	113	59.5	67	—	77	38	18.5
Zambia	190	19.4	113	70.0	80	—	57	26	14.7
Botswana	190	20.8	106	55.8	65	—	84	41	17.6
Papua New Guinea	190	20.5	125	65.8	74	—	65	32	16.9
Perak, (Malaysia)	190	21.4	118	62.1	70	—	72	35	17.8
Quebec, (canada)	190	23.3	98	51.6	55	Quebec govt share of RTS (%) 29.3	92	50	23.3
Western Australia	190	—	—	—	—	—	—	—	—
Gold Price: US$ 600/oz									
Sierra Leone	325	24.1	268	82.3	164	—	57	24	12.9
Tanzania	325	20.7	186	57.2	112	—	139	76	27.8
Zambia	325	29.9	215	66.2	129	—	110	59	24.1
Botswana	325	31.5	170	52.3	104	—	155	84	27.4
Papua New Guinea	325	29.4	214	65.8	128	—	111	60	25.1
Perak (Malaysia)	325	32.0	193	59.4	116	—	132	72	27.4
Quebec (Canada)	325	32.1	184	56.8	107	Quebec govt Share of RTS (%) 32.7	141	81	32.1
Western Australia	325	—	—	—	—	—	—	—	—

* Includes negative flows of investor (note that this situation is only hypothetical)

66   Options for Developing Countries in Mining

(ii) The government's share of the real total surplus (RTS) falls for Sierra Leone, Tanzania, Zambia, Botswana and Perak (Malaysia), but increases for Papua New Guinea and Quebec. This is principally due to profit-related taxes, the additional profits tax in the case of Papua New Guinea and the profit-related royalty for Quebec. (The detailed consideration of these is taken up in the next section.) It should be pointed out that the increased take principally benefits the Quebec government and not the Federal government, where the percentage of the Federal government take actually falls from about 25 per cent in the US$300 case to 23 per cent in the US$400 and US$600 cases.

(iii) Absolute investor take rises as well as the internal rate of return. If one assumes a 10 per cent real rate of return, at US$450, all regimes except Sierra Leone are acceptable, which is only so if a price of US$600 real is expected.

(iv) Investor IRR is lower than that of the project; this is due to the imposition of dividend withholding tax on all regimes except Quebec, where the investor IRR and project IRR are identical.

Due to the nature of capital allowances, the year-to-year project flows vary. The immediate write-offs in Zambia and Malaysia give low income tax receipts in the early years, but these rise as the allowances are used up. For instance, in the US$450 case, the government take for Perak (Malaysia) goes as low as 23 per cent in 1986 and 1987 but steadies at 57 per cent from 1989. For Zambia, the take is 34 per cent in 1986 and 1987 but steadies at 64 per cent from 1989. On the other hand, in regimes where the capital allowances are distributed evenly throughout the life of the projects, the tax receipts, as in the case of Botswana and Papua New Guinea, remain relatively steady throughout the life of the project.

Another interesting feature to note is the revenue-mix of the total government take. This is illustrated in Figure 4.2 which summarises the take from the various fiscal elements based on the gold price of US$450 an ounce. Quebec (Canada) gets 41 per cent of its total government take (Federal and Quebec) from the profit-based mining tax, while Papua New Guinea gets only 0.4 per cent of its take from royalties. For Zambia the mineral tax is 30 per cent of the total government take, for Perak (Malaysia) 10 per cent, for Sierra Leone 12 per cent, for Botswana 13 per cent and for Tanzania 5.2 per cent. The implication of this for Papua New Guinea is that, during a bad year, the government take would be very low, since over 99 per cent of its take is profit related. In the case of Quebec, where the mining tax is profit related, the percentage derived from this tax when the price is at US$300 is 38.7 per cent, still higher than all regimes, and 43 per cent when the price is US$600. This is due to the progressive nature of the Quebec Mining Tax (see Appendix 1(G)).

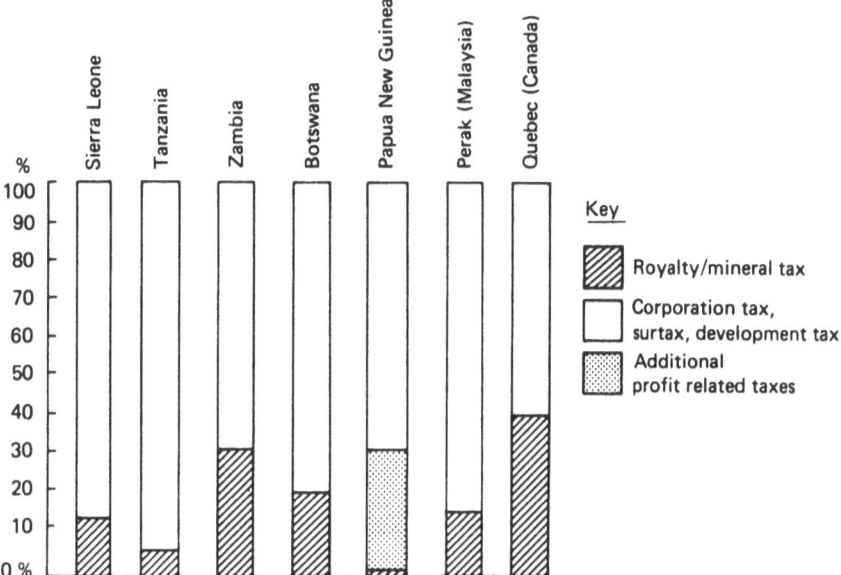

FIGURE 4.2 *Government revenue mix based on gold price of US$ 450/02 for case with 100% foreign investor equity participation*

The effect of discounting is also shown in the results. Those of Zambia, Botswana and Perak (Malaysia) show a positive real total surplus to the investor, but when the year-to-year flows are discounted by 5 per cent the figures become negative. If the net present value criterion is used to consider a project, that is, accept it if the NPV is positive, then the effect of discounting is to reject the project. The choice of discount rate is important. A 10 per cent real discount rate would have made the investors' net cash flow at the end of the life of the project for all regimes negative.

The project surplus (see Key Definitions) is essentially the surplus over and above the costs of the project, that is, capital cost, operating costs and fiscal dues. Table 4.2 summarises the relationship between revenue and project surplus for the three price scenarios for the 18-year planning horizon.

TABLE 4.2 *Project surplus as percentage of total revenue*

	US$300	US$450	US$600
Sierra Leone	16.3	37.5	41.0
Tanzania	38.6	47.3	48.6
Zambia	20.3	41.9	45.2
Botswana	39.4	55.3	58.0
Papua New Guinea	54.3	43.4	42.0
Perak (Malaysia)	34.5	47.8	49.7
Quebec (Canada)	62.9	48.4	43.2

The following observations can be made from Table 4.2:

(i) The percentage of the project surplus of total revenue increases as prices go up for all regimes except Quebec (Canada) and Papua New Guinea, which have schemes to cream off parts of the surpluses. In the case of Quebec, it is due to the progressive nature of the mining tax whereas for Papua New Guinea, it is due to the APT which triggers off when all capital costs have been recovered and the project earns a 13 per cent real rate of return (20 per cent nominal).

(ii) The year-to-year relationship between project surplus and revenue varies among the regimes. Countries with rapid write-off provisions for capital like Malaysia and Zambia have higher project surpluses in the early periods and lower surpluses in the later years. In the case of Sierra Leone and Botswana, they are relatively steady throughout the life of the project. For Papua New Guinea, the percentages of project surplus over revenue for the US$450 and US$600 cases range between 64 and 68 per cent in the early years but fall once the APT triggers off to 43 per cent and 42 per cent respectively. For Canada, because of the progressive nature of the mining tax, the project surplus falls steadily throughout the life of the mine, but never less than 40.6 per cent in the case of the US$450 price and 38 per cent for the US$600 price.

Case II – Results of Model with 51% Government Equity and 49% Foreign Investor Equity Fully Paid

Table 4.3 lists the results for all the regimes of the less developed countries for the three price scenarios. The real project IRR and that of the investor as well as the project surplus remain the same as the corresponding figures for Case I, since similar assumptions are being made. For the investor, while the IRR of his real net cash flows is unaffected, his absolute share of the RNCF falls because his equity is reduced from 100 per cent to 49 per cent.

An important aspect to note is that the return on government's fully paid equity, is, in fact, the same as the return to the project equity, government's share being 51 per cent of it. Hence governments planning to invest in the project will have to take into account the project viability in the context of their respective fiscal provisions. In making its decision, any government, apart from other criteria, will have to compare the rate of return on its equity with those of other investments. Another feature to note is that government's share of the real total surplus becomes higher when it participates because in addition to taxes and royalties, it receives dividends. Table 4.4 summarises the increase.

TABLE 4.3 *Difference in percentage government share of total surplus from case where it does not participate*

	US$300	US$450	US$600
Sierra Leone	—	+ 5.8	+ 9.1
Tanzania	+ 14.6	+ 20.5	+ 21.9
Zambia	+ 1.9	+ 15.3	+ 17.2
Botswana	+ 13.1	+ 22.6	+ 24.3
Papua New Guinea	+ 18.2	+ 17.4	+ 17.6
Perak (Malaysia)	+ 10.9	+ 19.5	+ 20.6

It can be observed that, generally, as the project becomes more profitable, that is, with higher prices, the share of government take goes up as a result of participation. The absolute increases vary among the different regimes, lower for Zambia and Sierra Leone compared to Botswana and Tanzania. For Papua New Guinea, the increase is relatively constant because the excess profits arising from higher prices have been taken by the government via the additional profits tax. The reason for this can be learnt from Figure 4.3 which shows the government revenue mix from the various fiscal elements and dividends as a result of taking 51 per cent equity in the project based on a gold price assumption of US$450.

It can be seen that dividends form as much as 45 per cent for Botswana's Government net cash flow, compared with 38 per cent for Tanzania, 37 per cent for Perak (Malaysia), 33 per cent for Papua New

TABLE 4.4 Results of model with 51% government equity and 49% foreign investor equity fully paid

	Real total surplus (RTS) (US$ (m))	Real project IRR (%)	Government				Investor		
			Share of RTS (US$ (m))	Share of RTS (%)	Share of discounted (5%) RTS (US$ (m))	Real IRR on fully paid equity (%)	Share of RTS (US$ (m))	Share of discounted (5%) RTS (US$ (m))	IRR on real net cash flows
Gold price: US$ 300/oz									
Sierra Leone	55	2.5	60*	100.0	31	2.5	−5	−9	0
Tanzania	55	6.7	47	85.5	22	6.7	8	0.1	5.1
Zambia	55	3.7	54	98.3	27	3.7	1	−4	0.7
Botswana	55	6.0	49	89.1	24	6.0	6	−1	3.8
Papua New Guinea	55	8.0	45	81.8	22	8.0	10	−1	5.7
Perak (Malaysia)	55	6.0	50	90.9	24	6.0	5	−1	3.6
Gold price: US$ 450/oz									
Sierra Leone	190	15.1	178	93.7	104	15.1	12	1	6.1
Tanzania	190	20.8	152	80.0	86	20.8	38	19	18.5
Zambia	190	19.4	162	85.3	93	19.4	28	13	14.7
Botswana	190	20.8	149	78.4	85	20.8	41	20	17.6
Papua New Guinea	190	20.5	158	83.2	90	20.5	32	15	16.9
Perak (Malaysia)	190	21.4	155	81.6	88	21.4	35	17	17.8
Gold price: US$ 600/oz									
Sierra Leone	325	24.1	297	91.4	176	24.1	28	12	12.9
Tanzania	325	30.7	257	79.1	151	30.7	68	37	27.8
Zambia	325	29.9	271	83.4	159	29.9	54	29	24.1
Botswana	325	31.5	249	76.6	147	31.5	76	41	27.4
Papua New Guinea	325	29.4	271	83.4	159	29.4	54	29	25.1
Perak (Malaysia)	325	32.0	260	80.0	152	32.0	65	35	27.4

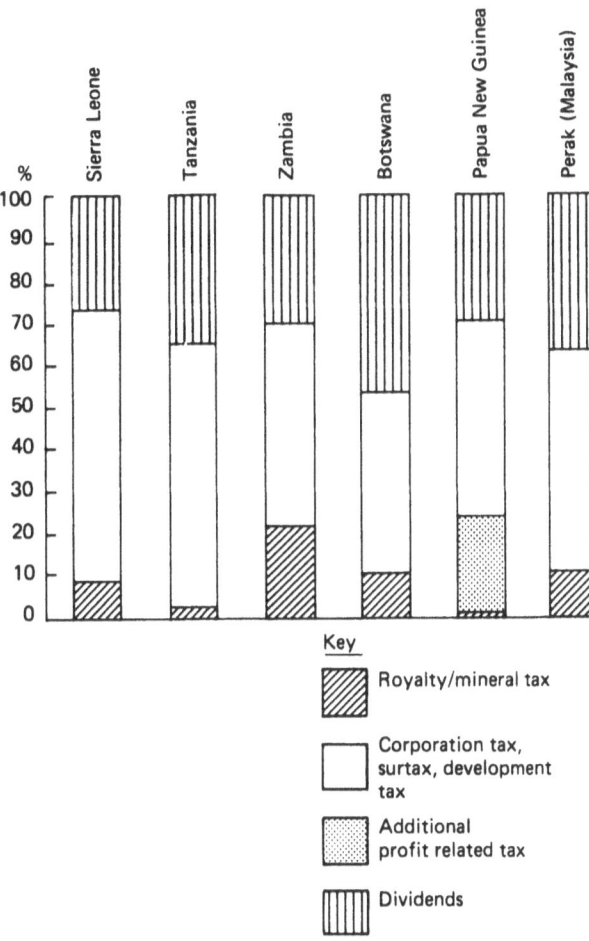

FIGURE 4.3 *Government revenue mix based on gold price of US$ 450/02 for case with 51% government fully paid equity*

Guinea, 34 per cent for Zambia and about 30 per cent for Sierra Leone. It should be noted that Botswana has a bigger project surplus than the others at the US$450 price level. For Botswana, Zambia and Perak (Malaysia), one reason for the higher rate from dividends when compared to Papua New Guinea is that in Papua New Guinea's case the rent resource tax reduces the amount available for declaring dividends.

It would, however, be more meaningful to look at the real IRR on fully paid equity, which essentially is the real project IRR. At the

US$300 price, the return is only 2.5 per cent for Sierra Leone compared to about 8 per cent for Papua New Guinea, which is the highest in the study. If government can achieve a higher return than this it makes sense for the government to invest elsewhere. However, at higher prices the return in government equity is as one would expect, healthy for all cases, ranging from 15 to 32 per cent real.

In financial terms, sovereignty can be implicitly measured as the economic rent derived from the exploitation of the resource. From the government's point of view this basically means all its receipts in terms of fiscal dues and dividends from its participation, minus all the expenditure in relation to investment and collection of the tax. Looking purely in terms of participation, the dividends it receives would be the return on its investment in the venture on its equity. If for the same amount of equity investment, government can make a better return elsewhere, then the opportunity cost of its funds may be high. In these circumstances it might make little sense to insist on majority ownership. In the above analysis, even in the best scenario, government take does not increase by more than 25 per cent compared to a situation where it does not participate.

A reason usually given for government ownership of shares is that it could obtain control rather than optimise investment. There are benefits in participation, the principal being direct experience and responsibility in mining. However, majority equity share does not imply control in all cases.

Case III – Results of Model where 31% of Government Equity is Fully Paid and 20% Free

Table 4.5 summarises the results for the three price scenarios. Looking at the project first, the real surplus falls from the corresponding two scenarios discussed above because of the deduction of interest on the loan. The difference is much less in the case of Sierra Leone and Papua New Guinea because of the interest withholding tax which reduces their interest rates by 45 per cent, that is, effectively to 6 per cent instead of 11 per cent, and this is 1 per cent below the rate of inflation. For Malaysia, which has no interest withholding tax, the real total surplus is therefore much lower than the others. To the extent that the actual amount of the loan is not index-linked, there is a gain in real terms to the project as repayment of the loan is made in later years, provided the interest rate is equal or below the inflation rate. In the model the interest rate is 4 per

TABLE 4.5 Results of model where 31% of government equity is fully paid and 20% free

	Real total surplus (RTS) (US$ (m))	Real project IRR (%)	Government					Investor		
			Share of RTS (US$ (m))	Share of RTS (%)	Share of discounted (5%) RTS (US$ (m))	Real IRR on fully paid equity (%)	Share of RTS (US$ (m))	Share of discounted (5%) RTS (US$ (m))	IRR on real net cash flows
Gold price: US$ 300/oz										
Sierra Leone	55	3.8	62*	100.0	36	3.8	−7	−9	0
Tanzania	52	9.4	49	94.2	26	9.4	3	−3	2.2
Zambia	53	4.4	56*	100.0	31	4.4	−3	−7	0
Botswana	52	8.0	50	96.2	28	8.0	2	−4	1.3
Papua New Guinea	55	10.8	50	90.9	28	10.8	5	−2	3.3
Perak (Malaysia)	51	7.9	50	98.0	27	7.9	1	−4	0.7
Gold price: US$ 450/oz										
Sierra Leone	190	23.2	180	94.7	108	23.2	11	2	6.4
Tanzania	187	33.7	154	82.4	90	33.7	33	16	18.5
Zambia	188	30.8	164	87.2	96	30.8	24	11	14.4
Botswana	187	31.7	150	80.2	89	31.7	37	17	17.6
Papua New Guinea	190	32.6	163	85.8	97	32.6	27	12	16.2
Perak (Malaysia)	186	34.3	155	83.3	90	34.3	31	15	17.6
Gold price: US$ 600/oz										
Sierra Leone	325	38.8	298	91.7	180	38.8	27	27	14.3
Tanzania	322	51.6	258	80.1	154	51.6	64	35	30.3
Zambia	323	50.4	272	84.2	163	50.4	51	27	26.0
Botswana	322	50.3	251	78.0	151	50.3	71	39	29.5
Papua New Guinea	325	49.0	276	84.9	166	49.0	49	26	26.3
Perak (Malaysia)	321	54.0	260	81.0	155	54.0	61	33	29.8

* Includes negative flows of investor. (Note that this situation is hypothetical only.)

cent above the inflation rate, and as such the burden of the loan is high. However, if discounting is considered there is a small benefit to the project as the real discount rate of 5 per cent is higher by one per cent than the real interest rate. Another factor that is important to the project is the net of tax interest rate which in this case ranges between 3.9 per cent and 5.5 per cent, depending on the tax rate of the regime in question. It can be noted that this rate is below the inflation rate of 7 per cent, and certainly makes financial sense to the project. It has to be mentioned that the interest rate chosen for the model is only to illustrate the issue, and the results hinge on the assumptions about the interest rate. Nevertheless, the general picture that emerges here is that the after tax rate of interest for most projects has been lower than the inflation rate. This results in higher project real IRRs for all the countries than the previous two scenarios for the identical project. Loan financing not only redistributes the risk of the project, but also affects the timing of the flows, which favour the project if the absolute amount of the loan repayment is itself not index-linked.

The investor IRR's are lower for the corresponding Case II figures because of the 20 per cent 'free' equity paid by the investor to the government. (It does not affect the investor's decision to invest for the US$450 and US$600 cases if one assumes a real rate of return expectation of 10 per cent, although for the US$300 case, the project looks financially unattractive, as for all the cases.) Accordingly, the corresponding share of the government take is higher than in the case where government fully finances its equity (see Table 4.6 for the changes. The increase ranges between 0.3 and 3.3 per cent for the US$450 and US$600 cases, but is much larger for the US$300 case, emphasising that the regimes are severe for the marginal fields). It has to be stressed that 'free' equity is not strictly 'free' in the sense that it is usually in payment, among other purposes, for other infrastructure, costs of which have not been taken into account in the calculations (see the first section of chapter 5).

Government return on its 31 per cent fully paid equity is similar to the project IRR for reasons explained in Case II.

Case IV – Results of Model with 31 % Government Equity Paid Out of Dividend and 20 % Free

Table 4.7 summarises the results for all the three price scenarios. The project surpluses and project IRRs remain the same as in Case III above

TABLE 4.6 Change in government take under different equity financing schemes

	(a) Gov't take of real surplus when equity is fully paid			(b) Change in gov't take (%) in case where 20% equity is 'free' and 31% fully paid compared to (a)			(c) Change in gov't take (%) where 31% equity is financed out of dividends and 20% 'free' compared to (a)			(d) Change in gov't take (%) where 31% of equity is financed out of dividends and 20% is 'free' compared to (b)		
	US$300	US$450	US$600	US$300	US$450	US$600	US$300	US$450	US$600	US$300	US$400	US$600
Sierra Leone	100.0	93.7	91.4	—	+1.0	+0.3	—	+1.1	−0.6	—	−2.1	−0.9
Tanzania	85.5	80.0	79.1	+8.7	+2.4	+0.1	+1.0	+0.2	+0.1	−7.7	−2.2	−0.9
Zambia	98.3	85.3	83.4	+14.7	+1.9	+0.8	+0.2	+0.2	+0.1	−1.9	−2.1	−0.9
Botswana	89.1	78.4	76.6	+19.6	+1.8	+1.4	+0.6	+0.2	+0.1	−7.7	−1.6	−1.3
Papua New Guinea	81.8	83.2	83.4	+7.5	+2.6	+1.6	+1.8	+1.0	+0.6	−7.3	−1.6	−0.9
Perak (Malaysia)	90.9	81.6	80.0	+18.0	+3.3	+1.0	+0.7	+0.4	−0.2	−7.8	−2.1	−1.2

TABLE 4.7 Results of model with 31% government equity paid out of dividend and 20% free

	Real total surplus (RTS) (US$ (m))	Real project IRR (%)	Government					Investor		
			Share of RTS (US$ (m))	Share of RTS (%)	Share of discounted (5%) RTS (US$ (m))	Real IRR on fully paid equity (%)	Share of RTS (US$ (m))	Share of discounted (5%) RTS (US$ (m))	IRR on real net cash flows
Gold price: US$ 300/oz										
Sierra Leone	55	3.8	57	100.0	33	—	−2	−7	0
Tanzania	52	9.4	45	86.5	24	—	7	0	5.1
Zambia	53	4.4	52	98.1	29	—	1	−4	0.9
Botswana	52	8.0	46	88.5	25	—	6	−1	3.9
Papua New Guinea	55	10.8	46	83.6	26	—	9	1	5.8
Perak (Malaysia)	51	7.9	46	90.2	24	—	5	−1	3.6
Gold price: US$ 450/oz										
Sierra Leone	190	23.2	176	92.6	105	—	14	4	9.3
Tanzania	187	33.7	150	80.2	87	—	37	19	22.3
Zambia	188	30.8	160	85.1	94	—	28	14	18.0
Botswana	187	31.7	147	78.6	86	—	40	20	20.8
Papua New Guinea	190	32.6	160	84.2	94	—	30	15	20.1
Perak (Malaysia)	186	34.3	151	81.2	87	—	35	18	21.4
Gold price: US$ 600/oz										
Sierra Leone	325	28.8	295	90.8	177	—	31	15	17.8
Tanzania	322	51.6	255	79.2	151	—	67	38	34.7
Zambia	323	50.4	269	83.3	160	—	54	30	30.4
Botswana	322	50.3	247	76.7	148	—	75	42	33.4
Papua New Guinea	325	49.0	273	84.0	163	—	53	29	30.9
Perak (Malaysia)	321	54.0	256	79.8	152	—	64	36	34.4

because the method of financing government equity does not affect the flows to the project.

There are some interesting observations as far as government take is concerned. First of all, it is marginally above Case II where government fully finances its equity because of the benefit from 'free' equity (see Table 4.7). However, when compared to the case where the remaining 31 per cent of the equity is fully paid, government take actually falls principally because the interest rate for the outstanding equity unpaid is 4 per cent above the inflation rate (that is, 11 per cent). Table 4.6 shows that for the US$450 and US$600 cases, the fall ranges between 0.9 and 2.2 per cent, whereas for the US$300 case, it ranges from 0 to 7.8 per cent.

There are some advantages of government financing its equity out of dividends. The first is that the project itself pays for the government finance, thus releasing money which would have gone from mining to other projects. Specifically, government funds that have been diverted from the mining project should at least earn not only the project IRR but also the interest paid by the government on borrowed funds for its equity. In some sense the project IRR plus the interest paid for the government borrowing for equity reflects the opportunity cost of government funds *vis à vis* the mining project. The interest rate at which government equity is financed thus becomes crucial in determining the over-all take, and in the choice of the method of equity financing. If the project is a lucrative one, that is, the US$600 case, the results show that it makes more sense for government to pay directly for its equity share instead of financing its share of the equity out of dividends, because of the high real interest rate which the government must pay.

It has to be stressed that this analysis only focused on selected policy issues regarding government financing of its equity. Project financing is a complex enterprise, and the range and sophistication of instruments that are published daily in the *Wall Street Journal*, the *Financial Times* and other business periodicals display the breadth of imagination being used by finance managers. The type of financing mechanism that will be used will depend on the stage of development of the country concerned and the sophistication of its financial know-how. Although the form of financing can sometimes change the economics of the project, ultimately it is the intrinsic profitability of the project that will produce the surplus. The method of financing purely redistributes the risks, it does not eliminate them; it affects the timing of financial flows, but not the ultimate economic rent.

Sensitivity Tests

To evaluate the effectiveness of some of the key parameters, sensitivity tests are performed on particular regimes for the US$450 price case. Sensitivity testing is basically a simulation analysis in which key variables are changed and the resulting change in the rate of return observed as well as the effect on other parameters such as government take. What is of more interest are not the absolute values but the *intensity* and *direction of change*. Absolute values hinge on assumptions. Sensitivity analysis, however, does not evaluate the risk associated with a mineral investment alternative. In order to measure risk, probability estimates are required for the possible variations in each of the input variables from their expected values. In fact, sensitivity analysis helps to define the 'strategic' variables which should be discussed in risk analysis. The results are presented in Table 4.8.

(A) *Alteration of Royalty Structure*

Given the relatively high mineral tax of Zambia, it was decided to see if its removal would improve the attractiveness of the marginal field. The results showed that for the US$300 case, even a fall in the government take by 21.9 per cent only gives the investor real IRR of 4.7 per cent (an improvement of 4 per cent), which falls short of the 10 or 12 per cent real rate of return considered normal in the industry. In any case, the investor's real discounted net cash flow remained negative. For the higher price scenarios, the investor benefits from the increased economic rent at the expense of government.

The royalty rate for Papua New Guinea was raised from $1\frac{1}{4}$ to 10 per cent of gross revenue. For the marginal field (US$300 case), this further deteriorated the economics of the project. The real project IRR fell from 8 to 4 per cent. However, for the high price scenarios, it improved government take by 5 to 6 per cent, but still gave the investor an IRR that is about the norm of the industry.

The other test was to see what happened to the Perak (Malaysia) regime when a Quebec-style profit-based progressive mineral tax replaced the 5 per cent royalty. The results showed that while this improved the investor's IRR by 1 per cent for the US$300 case, it was still not sufficient to entice the investor. However, for the higher price cases, the progressive tax improved government's take by 2 per cent for the US$450 case and 4 per cent for the US$600 case. The scheme is still

TABLE 4.8 Sensitivity tests of selected regimes and price scenarios

	Real total surplus (RTS) (US$ (m))	Real project IRR (%)	Government				Investor		
			Share of RTS (US$ (m))	Share of RTS (%)	Share of discontinued (5%) RTS (US$ (m))	Real IRR on fully paid equity (%)	Share of RTS (US$ (m))	Share of discontinued (5%) RTS (US$ (m))	IRR on real net cash flows
(1) *Zambia: Case 1* Remove mineral tax									
Gold price @ US$300/oz	55	8.1	41	74.5	23	–	14	–1	4.7
Gold price @ US$450/oz	190	23.3	115	60.5	68	–	75	37	18.2
Gold price @ US$600/oz	325	33.9	191	58.8	114	–	134	74	27.6
(2) *Papua New Guinea: Case 1* Increase royalty from 1 % to 10%									
Gold price @ US$300/oz	55	4.2	48	87.3	29	–	7	–7	2.1
Gold price @ US$450/oz	190	17.7	137	72.1	81	–	53	24	14.4
Gold price @ US$600/oz	325	26.4	231	71.1	138	–	94	50	22.3
(3) *Malaysia: Case 1* Replace 5% royalty by Quebec-style mineral tax (profit-based)									
Gold price @ US$300/oz	55	7.1	41	74.5	24	–	14	–1	4.6
Gold price @ US$450/oz	190	20.7	122	64.2	73	–	68	32	17.1
Gold price @ US$600/oz	325	30.0	207	63.7	124	–	118	64	25.6
(4) *Botswana: Case 1* Incorporate PNG APT scheme									
Gold price @ US$300/oz	55	6.0	42	76.4	25	–	13	–3	3.8
Gold price @ US$450/oz	190	18.7	130	68.4	77	–	60	28	15.4
Gold price @ US$600/oz	325	27.4	222	68.3	133	–	103	55	23.3

				Government				Investor		
	Real total surplus (RTS) (US$ (m))	Real project IRR (%)	Share of RTS (US$ (m))	Share of RTS (%)	Share of discounted (5%) RTS (US$ (m))	Real IRR on fully paid equity (%)	Share of RTS (US$ (m))	Share of discounted (5%) RTS (US$ (m))	IRR on real net cash flows
(5a) Papua New Guinea Case I Immediate capital write-off instead over life of mine									
Gold price @ US$300/oz	55	8.9	33	60.0	19	—	22	3	6.4
Gold price @ US$450/oz	190	22.1	125	65.8	72	—	65	33	18.2
Gold price @ US$600/oz	325	31.5	213	65.5	126	—	112	126	26.7
(5b) Papua New Guinea: Case I Increase income tax from 33 to 45%									
Gold price @ US$300/oz	55	6.4	41	74.5	25	—	14	−2	4.1
Gold price @ US$450/oz	190	18.7	129	67.9	77	—	61	28	15.4
Gold price @ US$600/oz	325	27.4	219	67.4	132	—	106	56	23.3
(6) Perak (Malaysia): Case IV Remove 'free' equity 51% of govt equity paid out of dividends									
Gold price @ US$300/oz	51	7.9	42	82.4	22	—	9	0.2	5.2
Gold price @ US$450/oz	186	34.3	149	80.1	86	—	37	20	23.6
Gold price @ US$600/oz	321	53.9	254	79.1	150	—	67	38	37.0

(7) *Sierra Leone: Case 1* Incorporate tax holiday for 5 years from commencement of production (i.e. no income tax and surtax)							
Gold price @ US$300/oz	55	6.2	59*	100.0	36*	—	−4	−13	0
Gold price @ US$450/oz	190	25.2	149	78.4	88	—	42	17	12.0
Gold price @ US$600/oz	325	39.7	237	72.9	141	—	88	47	22.3
(9) *Papua New Guinea: Case 1* Remove royalty and dividend withholdings tax Allow capital to be written off immediately							
Gold price @ US$300/oz	55	9.5	21	38.2	11	—	34	11	9.5
(10) *Sierra Leone: Case 1* Remove surtax, royalty, dividend withholding tax and allow for immediate capital write-off							
Gold price @ US$300/oz	55	8.1	28	50.9	15	—	27	7	8.1

not as robust as an APT from Federal government's point of view, but to the extent that royalties go to Perak State, the position is a better one.

The Papua New Guinea APT scheme was grafted onto the Botswana regime. For the US$300 case, APT was not payable because the real project IRR of 6 per cent was below the real accumulation rate of 13 per cent or 20 per cent nominal, that is, similar to that prescribed in Papua New Guinea. However, for the US$450 and US$600 cases, APT is payable and government take improved by 12.6 per cent for the US$450 case and by 16 per cent for the US$600 case.

(B) *Effect of a Change in Tax Depreciation Rate*

As discussed earlier, the various regimes under study have different tax depreciation schedules and therefore affect the timing of flows to the different parties. To test depreciation as a proxy for scale of project and its effect on the IRRs, accumulation for APT and impact on over-all government take, sensitivity testing was done on the Papua New Guinea capital allowance structure for Case I, that is, 100 per cent foreign investor equity. (It should be noted that Papua New Guinea is the only regime which has the APT scheme embedded in the legislation.) Immediate capital write-off was assumed instead of the declining balance depreciation scheme spread over the life of the mine. The effects are:

(i) The real project IRR increased marginally by 0.9 per cent for the US$300 case, 0.7 per cent for the US$450 case and 2.1 per cent for the US$600 case because of higher flows earlier in the life of the project compared to later ones.

(ii) The investor's real IRR increased by a similar magnitude. It should be noted again that the investor's real IRR is lower than that of the project because of the dividend withholding tax.

(iii) Government's real take fell by 3.6 per cent for the US$300 case because of the immediate write-off concession, holds steady for the US$450 case (fell in terms of its net present value) and fell by 0.3 per cent for the US$600 case. The reasons can be seen when one examines the returns from income tax and APT. Looking at income tax first, the respective total income tax paid throughout the life of the project remains the same, but the timing changes. For instance for the US$300 case, the effect of immediate capital write-off is to increase the losses of the project in 1986, the first year of the assessment, from US$2.3 million to US$30.42 million. The losses are carried forward in subsequent years. While the project is liable for an income tax of US$1.09 million in 1987 for the case where depreciation is over the life of the mine, no tax is payable in the case where the capital write-off is immediate. Indeed, no tax is payable till 1990, when the losses are fully written off. From 1991 onwards, however, the amount of tax paid is higher than the normal regime. A similar effect is seen for the US$450 and US$600 cases with tax payments delayed for two years for the US$450 cases, and one year for the US$600 cases. This means that faster capital write-off benefits the marginal project more than the highly profitable project.

Turning to the impact on APT, it can be seen that no APT was payable for the US$300 case, but the accumulated net assessable revenue in 2000 is US$121 million compared to US$160 million for the life-of-mine depreciation cases, because faster capital write-offs favour the project. The impact is better seen for the US$450 and US$600 cases where, as was noted previously, APT was payable (see Table 4.9).

TABLE 4.9 *Papua New Guinea: impact on additional profits tax of immediate capital write-offs*

	Capital write-off over life of mine		Immediate capital write-off		
Gold price per oz	(a) Total APT payable	(b) Year and amount when APT first paid	(c) Total APT payable	(d) % Increase from (a)	(e) Year when APT first paid
US$300	–	–	–	–	–
US$450	US$86.6 m	1991 (US$5.5) m	US$89.3 m	1.2	1990 (US$4.2 m)
US$600	US$158.4 m	1988 (US$0.9) m	US$159.6 m	0.8	1988 (US$5.1 m)

The table shows that receipts for the government from APT increase by 1.2 per cent and 0.8 per cent respectively for the US$450 and US$600 cases. For the US$450 case, the effect of a faster capital write-off is to bring forward by one year the payment of APT whereas for the US$600 case, although APT is first due in 1988 for both cases, a larger amount is payable when there are faster capital write-offs; US$5.1 million compared to US$0.9 million if capital write-off is over the life of the mine.

The impact of faster capital write-offs may be generalised as follows: *Faster capital write-offs do not affect in nominal terms the total income tax paid over the life of the project, but they affect of the timing of the flows to the respective parties. For the project, they enable higher flows in earlier years compared to later on and thereby increase the internal rate of return and net present value of the project; correspondingly, government net present value of its take falls because receipts from income tax are delayed. However, faster capital write-offs give rise to higher receipts from APT.*

As a policy guideline for government, it would seem that faster capital write-offs help the marginal project. For more prospective projects, giving faster capital write-offs favours the government where there is a rent resource tax or additional profits tax based on profitability. Where there is no additional profits tax and the project is a lucrative one, faster capital write-offs mainly favour the investor. In these circumstances, there might be a case for government asking the investor to guarantee reinvesting into the economy part of the capital base written off as suggested under Resource Rent Tax in chapter 3.

(C) *Increasing Income Tax Rate*

It was noted that Papua New Guinea's income tax rate of 33.3 per cent was low compared to those of the other developing countries. An increase to 45 per cent reduces the financial prospectivity of the US$300 field, but continues to make the other two price scenarios attractive. Government take is increased by 3 per cent, while investor's IRR falls by about $1\frac{1}{2}$ per cent for both cases. It should be noted that, in the Papua New Guinea case, the combined limit of income tax rate and APT rate works out to be 70 per cent. An increase in income tax rate merely reduces the APT rate, but to the extent that the income tax is paid when there are profits, whereas APT is paid only when a pre-determined IRR for the project is achieved, it would make better sense from government's viewpoint if the income tax were raised. In any case, if US companies are involved, US tax creditability rules allow for foreign tax paid creditable up to a rate of 48 per cent. Any tax below this rate merely flows into the coffers of the US Treasury, if remitted.

(D) Removal of 'Free' Equity

This was tested in the Perak (Malaysia) regime by removing the 20 per cent 'free' equity and assuming 51 per cent of government equity is paid out of dividends. The first effect is to increase the investor's IRR by 1.6 per cent for the US$300 case, 2.2 per cent for the US$450 case and 2.6 per cent for the US$600 case, that is, the higher the price, the greater the benefit to the investor. This also means that the US$300 case continues to remain marginal.

Government's take of the RTS for the three cases fell respectively by 7.8, 1.1 and 0.7 per cent, that is, the higher the price, the lower the fall. Compared to the case where 51 per cent of the equity is fully paid, the respective fall in take is 8.5, 1.6 and 0.9 per cent.

(E) Effect of a Tax Holiday

It was noted that Sierra Leone had granted five-year tax holiday to mining companies. The effect of this was tested in Case I by removing income tax and surtax. To simplify the analysis, it was assumed that capital allowances that were provided during the tax holiday could not be carried forward when the period was over. Royalty continued to be payable. The most important effect was that it improved the investor's IRR for all cases, but for the US$300 case the project continued to be unattractive. However, for the US$450 case, the project was more attractive, giving the investor a 12 per cent IRR instead of 6 per cent previously. If the dividend withholding tax had been removed, the investor's IRR would have been 25.2 per cent for the US$450 case and 39.7 per cent for the US$600 case.

(F) Making the Marginal Case More Attractive

Basically, all the regimes are severe on the US$300 (that is, the high cost/marginal) case. The relatively least severe among the developing countries was Papua New Guinea's regime, giving the investor an IRR of 5.7 per cent. When royalty and dividend withholding tax were removed, and the investor was allowed immediate capital write-off with all other parameters remaining intact, the investor's IRR rose to 9.5 per cent, almost similar to the Quebec case. This rate can get a project moving, if other conditions, that is, non-fiscal matters, are right.

For Sierra Leone, which appears to have the most severe legislative scheme, it will require the removal of surtax, royalty, dividend withholding tax and an immediate capital write-off to give the investor a real rate of return of 8.1 per cent.

This raises the issue of what base rate of return is acceptable to the project. No doubt it will depend on the objectives of the investor. If the investor is a multinational interested in cornering supplies or ensuring sufficient mineral raw material supply for its downstream activities, then it may be attracted to the project by a lower rate of return than when mining activity is an end in itself and profit-maximisation, the overriding objective. The former case is particularly well demonstrated in the case of bauxite where it is considered an initiation point for its ultimate conversion into alumina and subsequently aluminium.

Nevertheless, as mentioned earlier, there are risks, real or imagined, which the investor builds into the economic calculus. Such risk may include *political* risks, that is, liability to property rights, *geological* risks, that is, the prospect of finding a deposit of the right size, the right quality and with the right treatment characteristics, and *economic* risk such as the price of the mineral or the costs of production. The geological risk, in particular, is one that distinguishes mining from other sectors and, to accommodate this, the investor in practical terms will expect a higher rate of return. 'Political' and 'economic' risk are evaluated by both subjective and objective criteria by the particular investor, and the literature is well documented with methods by which multinationals evaluate such risks.

Historical trends of returns to mining show them to range between 4 and 20 per cent for USA between 1938 and 1961. For Canada, they averaged 9 per cent in recent years. But in these countries the multinational's perception of 'political' risk may be lower than for countries that are newly emerging and less familiar to the multinational. The multinational would expect to be compensated for the uncertainty in these countries in terms of higher rates of return. It has to be emphasised that such risks are never eliminated or reduced, only shared. The government of a developing country in a sense bears the perceived 'political' and other non-quantifiable risks of the multinational.

There are a host of other sensitivity tests that one could perform, but given the space, time and expense, the above are sufficient to explain the robustness of the regimes to changes in selected parameters and illuminate key policy issues. Such sensitivity testing is crucial at the time of the negotiations with the foreign investor to enable both parties to learn the financial effects of the fiscal package, as well as to examine government sovereignty, risk and opportunity costs in money terms.

5 Mineral Development Agreements: Variations from the Statutes

Despite the presence of elaborate financial and mining codes, contemporary mineral developments in less developed countries are effected by means of another instrument – the Mineral Development Agreement. All of the less developed countries in the study sample have concluded agreements with various mining companies to search for, exploit and market minerals found in their national borders. A mining agreement between the government and a multinational is a complex document which sets forth in detail the rights and obligations of the parties in the course of the various stages of the project. The legal basis for negotiation of the agreement is the mining code and other legislation relating to taxation, foreign exchange controls and imports, but government negotiators are usually given some freedom in negotiating provisions of the agreements to suit the prevailing circumstances.

This chapter focuses on the issues which have been isolated for special treatment in these agreements, while attempting to ascertain why these issues have been selected. It further comments on the utility of the various solutions, while focusing on the overriding concern of this work – that is, how sovereignty over resources can be expressed in practical terms. The chapter closes with a discussion on the negotiation process, and the relevance of an institutional framework as established by the mining and investment laws.

ISSUES IN MINERAL DEVELOPMENT AGREEMENTS

Form of Agreements

The issue of the form which a mineral agreement takes is often very controversial, and has been thought to have represented a significant impediment in the 1960s and 1970s to the development of the mineral sectors of many under-developed countries. The era of political independence brought with it the increasing awareness of the need to change former relationships so that economic self-reliance could accompany the political gains. It is within the context of this political momentum and its obvious concern about the forms of the institutional framework of the respective countries that greater emphasis was to be placed on the form of the relationship between the foreign investor and the state.

Unfortunately, in some of these countries, economic analysis, partly on account of an inadequate information base, did not keep pace with the political initiatives, so much so that some forms inconsistent with the structure of the specific industries were developed. Notwithstanding, significant progress was made in the sharpening of appreciation of the issues so that over time, more relevant choices in terms of form and content could be made by the respective states, taking into account their peculiar circumstances and the characteristics of the commodity in question. These developments also caused companies to reassess their roles, and countenance alternatives which were consistent to the changed and changing circumstances.

Mineral development in the less developed countries in the study, and, for that matter, in most of the third world, is accomplished principally by the co-operation of foreign capital and the local mineral resources. Through time, the forms of co-operation have changed from the traditional portfolio investment of rulers who sought to create a geological infrastructure in the respective colonies, to direct foreign investment on the part of private individuals and corporations attempting to fill the mineral needs of their respective home and other industries.

Foreign direct investment essentially was channelled to acquiring mineral rights from the host countries, and guarantees on the production and marketing of the minerals. The mineral rights were usually acquired by the grant of large concessions for extremely long periods, usually 99 years, as was evidenced in most former British colonies including Guyana, where bauxite leases were granted for such long periods. In more recent times, as shown by the examples in the study, even when a concession is granted, it is usually for much shorter periods (15 to 25 years) and for smaller areas. The recent mineral agreements concluded by the various states in the study group were either of the 'concession' or the 'joint-venture' types. The agreements of the pre–1970 period were invariably of the concession type, while all in the post–1970 period were subject to some form of joint development including the government or the local public/private sectors. It is significant that three of the four original concession style agreements recognised in the study have all been renegotiated to either effect local participation or make provision for it.

The Bauxite Mineral Prospecting and Mining Agreement 1961 between the Government of Sierra Leone, the Sierra Leone Ore and Metal Company and Aluminium-Industrie-Aktien-Gesellschaft was essentially a concession-type, but was renegotiated in 1976, and inter alia, made provision for local participation. Similarly, the Sierra Leone Rutile (Concession) Agreement of 1972 between the Government and Sierra Rutile Limited was renegotiated in 1975 to make provision for government participation (47 per cent). Iron development in Sierra Leone at Tonkolili and Marampa was subject to many agreements from as far back as 1937. However, in 1973 the agreement was modified to make provision for local participation and government representation on the board of directors.

Less developed countries, in order to gain a real stake in the actual development of the mineral resources, have taken up equity participation, and in some cases (notably Chile and later Guyana) outright ownership. The literature on the issues which led the various states to nationalise is quite voluminous, and can be referred to quite easily. However, this section deals with the joint-participation approach, which is the dominant trend.

In chapter 2, it was observed that joint participation with the foreign investor, even when not enshrined in statutory codes, was always a subject for specific policy pronouncements. The post–1970 mining statutes of Tanzania, Zambia, Botswana and Papua New Guinea have incorporated this concern and made the grant of mineral rights to foreign concerns conditional on the State having the right to participate in mineral development conducted by the foreign companies. In Sierra Leone and Malaysia investment guidelines for the natural resources sector indicate the necessity for local participation.

Table 5.1 lists some of the agreements entered into by the various states and it shows government equity participation of between 15 and 51 per cent in the mineral developments listed. On average there was a preference for 51 per cent government participation, invariably in a joint-production company registered locally. It should be pointed out that actual government equity participation in all the agreements cited was made after an ore deposit was established, and mining was contemplated. Governments did not as a rule contribute to on-going exploration programmes, or exercise the participation option during the

TABLE 5.1 Selected agreements and government participation

Country	Parties to Agreement Company	Mineral	Government participation (%) Total	'Free-equity'	Subscribed	Date of agreements	Comments
Tanzania	Uranerzbergbau Cmb (Uranerz)	Uranium	51	25	26	1979	Government only entitled to free equity if it takes up the remaining 26% of shareholding. Representation on board for at least 8 years; 4 govt nominees, 3 Uranerz nominees. Govt nominates chairman, Uranerz nominates general manager.
Zambia	(1) Roan Selection Trust Ltd (RST)	Copper	51	Nil	51	1969	Government bought 51% of assets of RST and AAC to the value of US$176m and US$118m. Entered into management, consultancy and sales contracts with the foreign firms. Representation on board 4 Zambians, 3 foreign reps. Zambian chairman, foreign managing director. In 1973 managing director also Zambian and AAC, RST lose veto power on board.
	(2) Anglo-American Corp. of Central Africa Ltd (AAC)	Copper	51	Nil	51	1969	
Botswana	(1) De Beers Botswana Mining Co. (Proprietary) Ltd	Diamond	50	15	35	1970	Amended 1972, then again in 1975. Equal number of rep. on board. De Beers responsible for technical supervision of mines.

Country	Company	Mineral			Year	Notes
	(2) Bamangwato Concession Ltd (BCL) and Botswana RST (Sales Ltd (BRST)	Nickel-Copper	15	—	1972	BCL is 85% owned by BRST which is itself owned 40% by the public, 30% by American Metal Climax (Amax), and 30% by the Anglo-American Corporation/Charter group (AAC).
Sierra Leone	(1) Sierra Leone Ore and Metal Co. (Seromco) (subsidiary of Swiss Aluminium Ltd – Alusuisse)	Bauxite	50	—	1972	Government and company each has 3 directors. Govt nominates chairman and Seromco the general manager. Seromco has right to manage for 10 years.
	(2) Sierra Rutile Ltd	Titanium	47	—	1972	Agreement renegotiated in 1975 to effect govt participation. Govt to nominate 2 directors.
	(3) Sierra Leone Selection Trust Ltd (SLST)	Diamonds	51	—	1970	Assets of SLST converted into joint venture with the national Diamond Mining Co. (DIMINCO). Govt entitled to appoint 6 directors and chairman. SLST entitled to appoint 5 directors and managing director.
	(4) The Sierra Leone Dev. Company (DELCO)	Iron Ore	—	—	—	Govt has the right to nominate 2 persons to the board of Directors of DELCO.
Papua New Guinea	Bouganville Copper Pty Ltd (subsidiary of Conzinc Rio Tinto Pty Ltd)	Copper	20	—	1967	Government has right to assign up to 25% of its shareholding to residents.

exploration phase of the project. In some of the cases, for example, Zambia, government participation took the form of purchase of equity in existing mines.

The same table also shows that some agreements have been effected by the grant to the government of 'free-equity', which Freeman (1980) more correctly calls 'equity issued as consideration for non-cash contributions to the venture' because, as the name implies, the government partner is bringing a concession, which is part of a long established geological infrastructure and a bundle of rights to the joint operation. Care should be exercised in interpreting the amount listed as 'free-equity' since different agreements have used various formulations in the structure of the joint company, which would indicate that some consideration is being given to government for making significant non-cash contributions.

Freeman also provides some useful distinctions in the numerical significance of the wording of the provisions which would confer 'free-equity' to government. In particular, he shows that:

(i) If the government is to achieve a prescribed level of 'free equity', say 20 per cent, and this is to be expressed as a percentage of the cost of the exploration programme, then the agreement should reflect that the fraction of $\frac{20}{80}$ (that is, expected 'free-equity'/remaining equity) should be used, that is, 25 per cent of the cost of the exploration programme should represent government's credit.

(ii) If the 20 per cent 'free-equity' is to be achieved by the rationalisation of the mining rights as a capital asset, and involves the issue of additional shares for it, then the depreciation base of the mining company is increased, taxable income is concomitantly reduced, and 'distributable profit of the mining company at the expense of tax-giving' is increased. It also upsets the debt-equity provisions, if controlled by statute or policy.

(iii) If the attempt is made to maintain the ratio of equity held by the respective partners, and also the debt/equity ratio, then an algebraic form can be derived which maintains that, but it is achievable only at the expense of increasing the loan component of the project.

For completeness, there are two other forms of mineral agreements, namely the management and service contracts, and the production-sharing agreements. These are popular in the oil industry, and have correlatives in the hard-rock mining industry. Management and service contracts have been noted in the mineral developments at Sar Chesmeh in Iran and San Isidro in Venezuela. This form of development is already prevalent in ventures between the state mining corporations and foreign companies, in which selected segments of the mineral development cycle have been contracted out to establishments that do not have an equity interest. Among those of note are the arrangements in the Dominican Republic where Rosario Resources Corporation has a major contract to supervise and advise the state corporation – Rosario Dominicana, in all aspects of its mining, and plant operations at Pueblo Viejo, and the agreement between the Guyana State Corporation – Guymine – and Green Construction Ltd for the stripping and mining operations at East Montgomery mine in Guyana.

The arrangements between a Russian mining corporation and the government of Guinea demonstrate most of the elements of a production-sharing arrangement. In this case, the Soviet Union provides all the financial, technical and managerial know-how to establish a bauxite-mining complex in Guinea in exchange for the delivery of a certain portion of the products to cover costs, while buying the remainder of the product on a prearranged basis, involving barter for consumables and other items from the Soviet Union. This latter form of development may become increasingly more important in bilateral state relations, especially in the context of the deteriorating liquidity position of less developed countries. It may not necessarily be the most efficient form in the use of funds, but it guarantees for the industrial country, markets for its industrial and other products in more stable long term arrangements.

Control and Management

Equity participation is usually accompanied by initiatives which permit the government to influence the direction and course of the joint-venture company. Invariably, the company is controlled by a board of directors, which reflects the shareholding of the respective partners (see Table 5.1). When government has 51 per cent ownership of the enterprise, it usually also has the right to the election of the chairman to the board. The foreign partner is generally allowed to nominate the general manager, who is answerable to the board, though in the case of Zambia this position was changed in 1973 when the government reserved the right to name both the chairman and managing director in the two operating state-controlled copper companies. The general manager is usually an ex-offico member of the board, and oversees the day-to-day operations of the company.

The general structure that has evolved is that the foreign partner is responsible for the initial management of the operations under the guidance of the board for some specified period such as ten years in the Alusuisse/Sierra Leone Agreement and eight years in the Tanzania/Uranerz Agreement. Minority rights are usually protected in cases of major issues, so much so that, in the initial Zambian participation arrangements in 1969, the minority partner had near total veto power. Though this was changed in 1973, the minority partner could dissent with board decisions by simply staying away from those meetings, since board assent required the presence of a quorum.

It is more common to have phase-in provisions for effecting the government carried interest over certain prescribed periods, as in the case of the Sierra Rutile/Sierra Leone Agreement where the government had the right to take up 47 per cent of the shareholding after eight years. Along with these phase-in provisions are the requirements for localisation of the management and other levels of the operations.

Malaysia has seen changes in ownership and control during the last 10–15 years, beginning first in the tin industry and subsequently being extended to other minerals. Until a major reorganisation in 1976, the Malaysian tin industry was essentially controlled by three transnationals – Anglo-Oriental (M) Senderian Berhad (wholly owned subsidiary of London Tin Corporation Ltd), Associated Mines (M) Senderian Berhad (owned by Charter Consolidated Ltd), and Osborne and Chappel Senderian Berhad. In 1976 the Malaysian Mining Corporation (MMC), 71.35 per cent owned by Pernas Securities (a national corporation) and 28.65 per cent by Charter Consolidated, assumed control of the London Tin Corporation Ltd and set up a joint-management company in January 1978 by the amalgamation of Anglo-Oriental and Associated Mines. MMC now controls about a quarter of Malaysian tin output, though dispersed (and therefore non-controlling) foreign holdings in its operating companies still account for about 55 per cent of the equity. See Thoburn (1981) and Govett and Robinson, 1980.

The important point to note is that equity control was achieved through normal western commercial methods with external participation still continuing, albeit with diminished control. Management and control of the industry are gradually phased out. This contrasts with outright nationalisation which reflects the belief that resource rents cannot be secured fully by taxation. All joint-ventures in Malaysia, be it with foreign companies or among various local entities (for example, state and federal), are viewed as business propositions and are dictated by market forces, although guided by national goals that require a minimum of 30 per cent participation by *bumiputra* (Malays), 40 per cent by other Malaysians and 30 per cent foreign ownership. The effect of this policy is that new ventures must be undertaken either in conjunction with Pernas Securities (federal corporation) or with various state corporations, which hold most of the bumiputra interests. Some states are not renewing mining licences to companies with less than 30 per cent bumiputra shareholders.

The Malaysian experience therefore reflects a delicate interplay of combining the objectives of redistribution, majority local ownership and long-term growth with continued external expertise and technology of an industry that had operated over several centuries by foreigners.

Fiscal Provisions

Despite elaborate statutory fiscal provisions, a common experience of all the countries studied is that mining development has generally been accompanied by significant variations, principally of a concessionary tenor, from those statutory provisions. Generally, the older the agreement, the greater the tendency towards significant deviations from the statutory norms, while the newer agreements approach more closely the provisions laid down by statute. The latter tendency is best represented by Tanzania, Papua New Guinea and Botswana, where attempts are being made to make companies operate along statutory lines. This also reflects greater development of specific sectoral policies, which more closely accommodate the peculiarities of the sector in the respective countries.

A perusal of Table 5.2 reveals that even for the same commodity, the fiscal arrangements in the 1961 and 1972 bauxite agreements concluded by Sierra Leone were quite different. This difference is symptomatic of the time of conclusion of the agreement, as is shown in the Sierra Leone case histories represented in Table 5.2. In the case of the bauxite agreements, total tax liability of the companies in the 1961 and 1972 agreements was limited to 50 and 52 per cent respectively, while in the Rutile agreement of 1972 and the diamond agreement of 1970, this liability was set at 50 and 70 per cent respectively.

Another significant element highlighted in the same Table is that most of the agreements which were concluded on very concessionary terms

TABLE 5.2 A comparison of fiscal arrangements in agreements and statutory provisions

Parties to agreement	Fiscal arrangements in agreement	Statutory provisions	Comments
(A) SIERRA LEONE 1. *Government of Sierra Leone and Sierra Leone Ore and Metal Co. Aluminium-Industrie-Aktien Gesellschaft (ALAG)*	*Royalty* 1^s 6^d per long ton (Not deductible for tax purposes) *Income Tax* (1) Rate 45% (2) 5 year tax holiday (3) Losses carried forward indefinitely in time (4) Pre-production exploration expenses written off at 10% on straight line basis (5) Post production exploration expenses set off directly against income (6) Capital allowances: 5% or *output* reserves (7) Tax plus royalty liability of Co. is limited to 50% of Co.'s assessable income except that full royalty must be paid *Withholding tax* Dividends not subject to withholding tax except if double tax relief is possible *Other* (1) No customs duties for first five years of project (2) No export duties	*Royalty* Not available *Income Tax* (1) Rate 45% (2) Tax holiday may be granted (3) Pre-production exploration expense: 5% initial (4) Annual allowance. Greater of 5% or *output* reserves. (5) 15% Surtax on income tax payable *Withholding Tax* 45% on dividends and interest *Customs duties prescribed:*	1961 agreement for development of bauxite agreement made provision for change of financial arrangement after 15 years. This agreement was subsequently renegotiated, and inter alia participation was catered for

2. Sierra Leone Ore and Metal Co. Ltd (Seromco) (a subsidiary of Swiss Aluminium Ltd. – Alusuisse)

Royalty
Leones 55 per ton. (Deductible for tax purposes)

Income Tax
(1) Joint Venture Co.: 45%
(2) Joint Venture Co.: total tax liability 52%
(3) Alusuisse only to be taxed on 10% of its engineering fees and 30% of its management fees
(4) Pepel Alumina Co. given 5 year tax holiday

Withholding Tax
No withholding tax

Others
(1) No customs duties for 5 years
(2) Import levies not to exceed 5% for 5 years

Royalty
1972 agreement for exploration and development and mining of bauxite in a 50/50 joint-venture.

Income Tax
As above

It also included the establishment of an Alumina plant to be managed by a wholly owned subsidiary of Alusuisse – Pepel Alumina Co. Fees of 10% of total investment cost and 2.5% per annum of sales had to be paid to Alusuisse for engineering and other services during construction stage, and for acting as managing partner respectively.

TABLE 5.2 (Continued)

Parties to agreement	Fiscal arrangements in agreement	Statutory provisions	Comments
3. Sierra Rutile Ltd	*Royalty* Rutile (with $TiO_2 > 95\%$) Le 3/tonne for $< 25\,000$ tonnes Le 5/tonne for $> 25\,000$ tonnes $< 50\,000$ tonnes Le 6/tonne for $> 50\,000$ tonnes Ilmenite (with $TiO_2 > 50\%$) 10c per tonne (Royalty treated as tax credit) *Income Taxes* (1) Statutory tax and surtax applicable (2) Total tax liability limited to 50% of profits (3) No limitation can reduce tax liability below level of royalty due (4) 5 year tax holiday *Withholding Tax* Company exempted *Other* (1) Customs duties payable	*Royalty* Ilmenite 10c per tonne	1972 agreement for exploration, development and mining of titanium minerals. This agreement was renegotiated in 1975 with the following terms: (1) Tax holiday reduced to 3 years from production day (2) Withholding tax imposed on dividends paid or due during tax holiday period (3) After 8 years from production day, government can hold 47% of shares, and shareholders liable to a 10% withholding tax on dividends (4) For a period of 5 years after tax holiday, company shall pay 50% of customs duties on fuel other than petrol and kerosene up to maximum of Le 100 000. After that period, further duties subject to negotiation (5) Up to 8 years after the production day, the company can hold funds in any account, and in any country. After this period, co. must comply with exchange control provisions (6) Govt has right to nominate 2 directors on the board from production day

4. **Sierra Leone Selection Trust Ltd (SLST)**

Royalty
5% ad volorem exempted tax liability is greater than 70% of profits

Income Taxes (As above) plus diamond industry Profits tax: $27\frac{1}{2}$% of profits
(1) DOMINCO to be liable to statutory tax, diamond industry tax, and other taxes on profits provided that the total liability of DOMINCO in respect of these taxes is equal to 70% of such profits
(2) If above is met, then company exempt from all rents, taxes, royalties, fees, export duties charges or impositions of any kind

1970 agreement to create the National Diamond Mining Co. (Sierra Leone) Ltd (DIMINCO) SLST operated concessions since 1935, and government bought 51% of equity with negotiable government bonds

(B) TANZANIA

1. **Uranerzbergbau GmbH (Uranerz)**

Royalty
3% ad valorem (deductible for tax purposes)

Income Tax
(1) Rate of $22\frac{1}{2}$% for first 4 years
(2) Rate of 50% for fifth year
(3) Depreciation allowances: 40:10:10:10:10:
(4) Additional profits tax on surplus of return of 20% and 25% respectively 40% and 50% after attainment of

Withholding Tax
Dividends 10%
Interest $12\frac{1}{2}$%

Others
Imports duty free

Royalty
For uranium 5% ad valorem

Income Tax
(1) General company rate 50%
(2) General foreign co. rate 55%
(3) Depreciation allowances: 40:10:10:10:10:10:

Withholding Tax
Dividends 10%
Interest $12\frac{1}{2}$%
Management & professional fees 20%
20%

1979 agreement for exploration development and mining of uranium

TABLE 5.2 (Continued)

Parties to agreement	Fiscal arrangements in agreement	Statutory provisions	Comments
(C)ZAMBIA 1. Anglo-American Corporation (AAC)	*Royalty* As per statue *Income Tax* (1) As per statute, but total tax liability not to exceed 73.05% (2) All payments on bonds exempt from tax	*Royalty* (1) No royalty chargeable after 1976. Replaced by Mineral Tax (2) Before 1976, rate was 13.5% of price of Cu less K16 *Income Tax* (1) Rate of 45% (2) Depreciation allowance for mines after 1953. All capital expenditure expensed against current income (3) Mineral Tax: 51% (Deductible in computing income tax)	1969 agreement for government to participate 51% in the copper industry. Similar agreement concluded with Roan Selection Trust (RST) which, like AAC, retain a 20% beneficial interest. Govt through state corporation – Mendeco acquired 51% interest in operating company.

Management & Consultancy Contracts (AAC)
Payments of:
(1) ¾ of 1% of gross sales
(2) 2% of profits before tax
(3) 3% engineering service fee based on construction cost
(4) 15% of first year emoluments for expatriates as a recruitment fee

Sales and Marketing Contracts
(1) ¾ of 1% of gross copper sales
(2) 2½% of cobalt sales

Withholding tax
Dividends 20%
Interest 10%

Others
(1) Export tax of 40% when price of Cu is above K 600 per long ton. Below this price, no tax payable

In 1973, review of agreement effected. This included:
(1) Redemption of all bonds and loans to AAC and RST
(2) All legislative guarantees revoked
(3) Capital expenditures written off as per statute
(4) Dividends subject to 20% withholding tax, and only 50% of after-tax-dividends remittable
(5) Termination of sales, marketing and management provisions by paying US$ 46m to AAC and US$ 34m to RST
(6) Government set up marketing agency to handle marketing of copper

102

TABLE 5.2 (Continued)

Parties to agreement	Fiscal arrangements in agreement	Statutory provisions	Comments
(D) BOTSWANA			
1. DeBeers Botswana Mining Company (Proprietary) Ltd	*Royalty* (1) On diamonds 5% ad valorem (2) Diamond levy ($\frac{1}{4}$ realised profit) to be dropped *Income Tax* (1) As per statute: 35% (2) Profits tax: 10% (3) Immediate expensing of capital expenditure *Withholding Tax* As per statute	*Royalty* (1) Precious stones: 10% ad valorem (2) Diamond levy: $\frac{3}{4}$ realised profit (both deductable for tax purposes) *Income Tax* (1) Rate of 35% (2) Capital allowances written off over life of mine (Mine life not to exceed 30 years) *Withholding Tax* Dividends 15% Interest 15%	1970 agreement, amended in 1972 for the mining of diamonds at Orapa. In 1975, when the Leblhakana mine was brought into production, agreement was amended to: (1) Introduce a variable royalty for different mines (2) Eliminate the 10% profits tax (3) Increase government equity from 15% to 50% and have equal participation on board (4) Indicate that any future statutory changes in tax rate will apply to company
2. Bamangwato Concessions Ltd (BCL) Botswana RST Ltd BCL (Sales) Ltd	*Royalty* (1) $7\frac{1}{2}$ of net income with an advance of Rand 750000 *Income Tax* (1) Rate of 40% (2) Tax rate increased by 1% for every 1% by which operating profit margin exceeds 48.5% to maximum rate of 65% *Withholding Tax* Waived	*Royalty* Precious metals 5% ad valorem *Income Tax* As abofe *Withholding Tax* As above	1972 agreement for the development and mining of copper and nickel at Selibi-Pikwe. Because of serious production difficulties government subsequently (1) Reduced royalty from $7\frac{1}{2}$ to 3% (2) Caused a restructuring of the sales agreement

PAPUA NEW GUINEA

1. Bougainville Copper Pty Ltd (Subsidiary of Conzinc Rio Tinto Pty Ltd)

Royalty

Income Tax

(1) Rate at 25% until investment recouped.
(2) Rate increased to 50% up to fifth year and remains until 26th year.
(3) After 26th year, rate increases by 1% in each of the following years until maximum of 66% (i.e. 62% effective tax is achieved).
(4) 20% of taxable deducted on allowance.
(5) Tax holiday for 3 years.
(6) Carry forward of losses indefinitely.
(7) Rapid depreciation.

Withholding Tax
Dividends 15%

Other
No customs duty for 10 years after completion of production, except comparable products are found locally.

Royalty
1¼% f.o.b. value

Income Tax
(1) For non-resident Co.: 45%
(2) For resident Co.: 33%
(3) Capital allowances
   Residual capital expnditure Remaining life of mine
   (a) For capital items: max. of 10%
   (b) For exploration: max. of 20%
   (c) Initial deduction: 25% of capital expenditure
   (d) Additional profits tax of (70-n) of surplus once prescribed rate of return of 20% orreduced prime interest of US plus 12%

Withholding Tax
Dividends 15%
Interest 45%

1967 agreement for the exploration, development and mining of copper. In 1974 agreement was renegotiated to give effect to the following:

(1) Tax exempt period ceased on 31 December 1973
(2) The deduction of 20% of taxable income in computation of the base was discontinued from 1st January of 1974.
(3) Company to pay normal tax as contained in statutes, and additional profits tax on any surplus
(4) Depreciation allowance orreduced
(5) Company instead of government required in 1967 agreement, required to meet all infrastructural costs for access mine
(6) Company to pay 50%c tonne of contained copper to Bouganville non-renewable resources fund
(7) Company to pay statutory import and stamp duties.
(8) Dividend withholding tax 15%
(9) Studies on forward processing to be done by company
(10) Conditions to be reviewed after 7 years

were subsequently renegotiated, with the principal change being the imposition of greater levies. The 1961 bauxite agreement, the 1972 Rutile agreement of Sierra Leone, the 1967 Bouganville copper agreement in Papua New Guinea, the 1970 diamond agreement of Botswana, are examples of agreements which were all renegotiated. In 1975 renegotiation of the Rutile agreement was significant since it attempted to enforce most of the statutory tax, customs and exchange control requirements, while removing the very liberal tax holiday period which was granted. Similarly, the 1969 Zambian copper agreements were renegotiated in 1973 to have applicable some statutory provisions, while also freeing the government of some 'patently uneconomic' arrangements which it concluded with its partners. However, it is also essential to note that governments, as in the case of the Botswana Selibi-Pikwe development, can be called upon to give greater concessions in order that a project can be kept alive.

With regard to the actual fiscal instruments, some general points emerged here. While tax holidays for long periods were at one time quite fashionable, contemporary mining development agreements reflect a shift away from this position. Fast capital write-off is the more prevalent method of allowing the investor quick recoupment of his investment. There has also been a tendency for all pre-production expenses to be recouped in such a way that the state shares in any surplus generated in the early years of the project. The other significant point to note is the preference in some countries (for example, Tanzania and Papua New Guinea) for a resource rent tax which was discussed in chapter 3. Also to be noted is that customs duties were generally waived, especially during the exploration phase of the project.

Non-Fiscal Issues

Pricing and marketing usually occupy a significant place in most modern mineral development agreements. For the government, the issues at stake are the need to obtain the highest prices and to have an impact on the marketing strategy of the enterprise. In particular, the agreement would make special reference to the necessity for arms-length trading, and may deal at length with questions of definition of affiliates and the manner in which affiliates could conduct business. The obvious concern is to limit the ability of the investor to transfer prices, thus reducing his tax liability in the jurisdiction of the host country, especially if it has high rates of marginal tax. The state may also have some concerns about the sale of products to certain countries, such as South Africa, or for some specific end use, and it may therefore seek these guarantees in the mining agreement.

Repatriation of funds to the home of the investor is a significant consideration in mining agreements. To meet this concern, the investor is given the privilege of maintaining an account either externally or internally, denominated in foreign currency. Within certain prescribed rules, the investor can make payments and receive monies in that account, but he can also be obliged to channel excesses above a stipulated amount into the mainstream of the operations of the central bank.

In chapter 2 it was observed that there was a perceptible evolution in the increasing economic development orientation of the newer mineral statutes. Similarly, this trend is to be observed in the mineral agreements. Specifically the agreement would oblige the investor to transfer technology, train locals, develop infrastructure, create local linkages, reinvest periodically, and to integrate his operations in the larger economy.

The agreements are usually governed by the local laws, but they generally make provision for the resolution of disputes between the host country and the investor in international courts or according to some rules of international law. This trend is typical of the Commonwealth countries, and is a marked departure from the practice in most Latin American countries where the *Calvo principle* has obviously taken hold. According to this principle, disputes between the investor and the state cannot be resolved in any international court, but rather must be dealt with only by the application of local laws. As a general rule, there is an attempt to preserve the legal principle of the sanctity of contract, though most would concede that this sanctity must be construed within the bounds of sovereignty and the inherent ability of the state to exercise its legislative competence in ways which best achieve their respective objectives.

WHY HAVE MINING DEVELOPMENT AGREEMENTS BEEN NECESSARY?

The preceding sections have indicated that despite the presence of elaborate national laws, large scale mineral developments have traditionally been accomplished by agreements which made exceptions to those laws. Obviously there is always the consideration that the body of statutes can be made obsolete with time, as the current and projected future realities dictate that other norms must be established. However, it appears that mineral developments are particularly prone to exhibit this tendency towards greater volatility in the instruments which facilitate mining. This section attempts to isolate the reasons for this phenomenon, as well as to identify those measures which may be put in place to facilitate the pace of development. Basically two issues are identified, and they relate specifically to the characteristics of mining and the structure of the mining industry.

Issues Inherent in the Characteristics of Mining

Mineral development is basically capital-intensive in nature and technology is in the hands of developed countries. It is only in the case of artisanal mining that the input factor proportions would show a greater use of labour, which typically is only minimally absorbed in the mechanised mining industry – the principal provider of the mineral inputs required in the industrialisation process.

Some obvious consequences which flow from the capital intensive nature of mining are:

a) If the country which seeks to develop the mineral resources is deficient in capital, it must first of all contemplate a role for the foreign investor in the development of those resources. The role assigned to the foreign investor will be reflected in the body of fiscal and non-fiscal regulations which the state considers essential for the attraction and retention of the investor within its boundaries. The central point to be borne in mind is that since the capital must originate from without, and since others are also competing for those resources, it is incumbent on the state to create the institutional framework which would facilitate the movement of the capital. Mining agreements reflect in part an attempt to create the institutional framework.

b) High capital intensity normally reflects itself in high productivity, which is usually the basis for establishing wages. Mining areas are therefore high productivity, high wage areas which in normal circumstances would be out of line with wage levels established in a dominantly labour-surplus economy. This is not a point to be treated lightly by governments which are mindful of developing their mining sectors, since the unconditional transplant of wage levels achieved in the mining sector to other secotrs of the economy where productivity is lower, can lead to

Mineral Development Agreements 107

economic dislocations in the economy. The need for a sensible wages policy becomes greater, as one introduces or increases mining activity in the country.

Mines are usually developed in remote areas where there is a serious lack of any physical and social infrastructure. A characteristic of mining therefore is that it usually makes tremendous demands on the provision of infrastructure, which would reduce the impact of its physical isolation. This physical isolation, combined with the productivity/wage concerns mentioned above, contributes significantly to the tendency for mining communities to develop into economic and social enclaves in the country. The majority of the developmental concerns which have been either encoded in statute or become the subject of mineral agreements are principally aimed at the integration of the mining operation in the framework of the total economy.

The period from the initial exploration through to the development of a mine can be quite long, with each succeeding stage making proportionately larger demands on resources. As a consequence, investors have to tie up large amounts of funds before they can entertain the prospect of recovering them. It is only within the framework of an assessment of the definiteness of the property rights which the state conveys will the investment be made. Mining agreements have therefore to define clearly the bundle of rights which the concession conveys. To this end, it is not only the presence of the statutes which matters, but also specific assurances, embodied in contract, which the investor needs. Ultimately, even in the context of relatively stable local laws, and an independent judiciary, the investor will seek final redress to some international forum, should disputes arise between itself and the state. As a further safeguard, and at the same time taking full cognisance of the depreciating value of money through time, the investor will seek mechanisms to recoup his investment as early as possible, whether through the vehicle of tax holidays, or other accounting considerations, be they accelerated write-off or special allowances.

Issues Inherent in the Structure of the Mining Industry

One of the distinguishing elements of the mining industry is that the distribution of the resource owners and the users is different, while the mining industry is typically controlled by few large vertically integrated mining companies. Obviously, this is not necessarily true across all mineral commodities, though in value terms, the generalisation is quite valid (see Bosson and Varon, 1977). In the last section the point was made that as a consequence of the capital-intensive nature of mining, mineral sector development in the capital deficient developing countries would of necessity presuppose that capital must be attracted. However, the other side of the coin indicates that since the loci of consumption and production are different, the end-users of mineral products must take capital to the immobile resource endowment so that useful products can be developed. Similarly, since the various entities need access to the resources, they are therefore also in a form of competition. To be able to go to the next step and decide whether it is an investor's or host country's market is more difficult since this is conditional on the type of mineral and its own characteristics.

The investor's concern about his organisation's ability to assure continuity of supplies and its longevity, invariably gets itself reflected in attempts at resource cornering, with levels of resource inventory which outstrip the present and future planned productive capacities of the organisations. Since 1950 the United States has built up its stockpile of strategic minerals, particularly tin, aluminium, cobalt, chromium and manganese, all of which the country has to purchase exclusively or mainly from abroad. Britain, France, Japan, Spain, Sweden, South Korea and Italy all operate stockpiles or are considering them (see Marsh, 1983).

Host countries, on the other hand, have a greater preference for immediate production, and therefore find it necessary to limit available sizes of concessions, while insisting on vigorous work programmes in their statutes or agreements. Similarly, host governments are recognising the need to play an active role in marketing as was seen in Zambia's experience. Some of them have become involved in discussions to provide investment insurance and other guarantees, whether on a bilateral arrangement or through influence in multilateral bodies.

Fundamental to the movement of capital resources from the industrial centres to developing countries is the question of the total tax burden which the company must accommodate. It is therefore crucial that developing countries should familiarise themselves with the tax treatment which the company will experience in his home country, since even if generous incentives are given by the host country, these may simply be taxed away by the investor's home country. Tax creditability at the investor's home base is a crucial element in contemporary mining agreements and it is one of the few occasions when the interests of the host and company can be merged to let each gain at the expense of a tax opportunity on the part of the investor's home country.

One of the most insidious and most addressed concerns which the host country has with respect to the operations of the integrated conglomerate is the latter's ability to shift cost, price and profitability centres among its subsidiaries regardless of where they are. Though this issue has been commented on by many observers, the phenomenon of transfer-pricing is believed to be a routine form of business for the transnational corporation, and it requires very sophisticated machinery to handle it. Though mining contracts usually attempt to build in safeguards against the use of transfer prices, it can only be effectively countered by developing strong capabilities in the taxing and other relevant institutions in the host country.

Developing countries have the long-term objective of becoming involved in the downstream end of mineral activity where the value added becomes greatest. Agreements and statutes often attempt to deal with this issue since it is one way in which the state can achieve optimum returns from the exploitation of the resource (for further discussion on integration, see UNIDO, 1978, and Radetzki, 1977).

NEGOTIATION OF MINING AGREEMENTS

Mining development agreements would continue to be a most important element in the promotion and conduct of mineral activities in developing countries. Over time, mining and other relevant statutes will be developed to lend expression to the various concerns of the investor and the host country. Though intrinsically desirable, it is best for the statutes to admit some flexibility as well as stability, since too many variables are at work in dynamic circumstances.

From the standpoint of the negotiating process, it is best for the state to have a comprehensive set of laws and guidelines to use as its starting point in discussion on mineral development. It is generally accepted that it is more difficult to negotiate away the existing laws, and when called on to do so, the state is in a greater position to exact a reasonable price.

The company's negotiating team is usually well prepared and in command of a greater body of facts concerning the development in question. However, this is no reason for the government team to resign in awe. Where it recognises that it cannot handle the discussions on its own, it could seek help from agencies such as the Technical Assistance Group of the Commonwealth Fund for Technical Cooperation and the UN Department for Technical Cooperation and Development, which are specialised in such matters. Secondly, but more importantly, the host country must be prepared to do its homework, since knowledge of the company, the commodity, the markets, and what others are doing, are the essential points around which the negotiation will take place, and respect developed across the negotiation table. Lipton (1976) has given a fairly comprehensive insight on the negotiation process.

It is the division of the economic rent which the state should see as its principal objective. Every tool, including the use of the computer, must be thrown into the desire to protect the economic bottom line. Since economic calculations are at the heart of the investor's decision criteria, the developing country must learn to manipulate the economic instruments illustrated in chapter 4. It must attempt to gain from the investor the assumptions which inform their judgement, and find suitable counters for the basis on which it sees the economic calculus unfolding. Sovereignty should then best be construed in terms of the state's ability to manipulate the conversion of the nation's resources into growth and developmental priorities.

A word of caution is necessary if it is to be interpreted that mastery at the negotiation table is all that is required to give the state true expressions of sovereignty. In fact, the negotiation process is just the beginning, and perhaps the easiest part of the whole project cycle. The state must develop its own capability through its tax authorities, mining administration, training and business acumen, to first monitor the activities of the company and then, more importantly, to engage in true partnership, as among equals. Mining is a multi-generational commitment, where there is no room for myopia.

6 Summary and Conclusions

The developing country desiring to initiate or increase the role of mining in the national economy is confronted with a range of complex issues which must be carefully handled. Foremost among them is the notion, based on geological observations, that the country has a physical endowment of minerals which can be commercially exploited. The first responsibility of the State, therefore, is to create the geological infrastructure which, in the initial stages, can simply be a regional mapping survey. To accomplish this objective, the State must make financial provision for this basic work, which can also be assisted by aid from the numerous multilateral agencies such as the United Nations Development Programme and existing bilateral aid programmes.

The indicated physical endowment requires a large commitment of funds and technology to convert the mineral resources into marketable products. This financial resource is unavailable in the typical less developed country and it must perforce come from external sources. Where the capital has an external origin, it carries with it the necessity for numerous considerations which involve preconditions for its attraction, its protection, and finally its repatriation. It should also be borne in mind that even when the capital is of local origin, there are a number of similar concerns such as stability of property rights and a fair economic return which have to be guaranteed before the investment will be made.

Chapter 2 focused on the institutional preparation for expressions of sovereignty by the state over its mineral resources. It isolated the issues, and the solutions which the state must take to monitor, regulate and control the form which mineral development should take. The review of the mining laws of the various countries illustrated that despite their socio-political and cultural differences, the two basic concerns of contemporary mining legislation are the need to regulate the sector and to make mining contribute to national development. The observation was made that there has been a clear evolution from the original emphasis of states to purely police the sector to a later concern about using the mining sector as a vehicle to achieve economic developmental goals.

The further point was made that the mining legislation should admit some flexibility which would allow it to react more precisely to the peculiar circumstances of the state. It was recognised that despite the similarities in the statutes of the various regions, each had variations that took account of their peculiar physical and human endowment, size and state of development. There are no hard and fast rules, but much can be gained by studying what others have done in similar circumstances.

Salient points of detail which have emerged in the study of the respective countries involve the attempts by states to pursue as follows:

(i) Differentiate and recognise areas of mineral development which are restricted to nationals (these areas are usually the low-technology, labour-intensive (ones);
(ii) require that foreign investment takes on a local corporate form;
(iii) promote the need for a vigorous work programme by entities which are technically and financially capable. To achieve the efficient use of the state lands it is not uncommon for work performance guarantees to be required of the investor, while periodic relinquishment of lands and or escalating rentals may be necessary during the life of the exploration licences.

Generally, succession in mining rights is usually guaranteed as long as the preconditions have been met. These preconditions have provided for work commitments, accurate and timely reporting, satisfactory labour practices including employment of local labour, local procurement of materials and supplies where applicable, and general adherence by the investor to the laws of the country.

The mining statutes generally make allowance for the resolution of disputes at three levels. In the first case, disputes between two licencees can be heard and tried by the chief executive established by statute. The chief executive decisions can be appealed to the Minister, whose decision may be final on a number of issues, though principally on the issue of award of mining rights. Alternatively, the decisions of the chief executive, especially on an interpretation of the law, can be challenged in the High Court. In some cases, even the decision of the Minister on the interpretation of the law can be challenged in court. It is to be noted that in the developed countries in the study, the practice is generally for all decisions of the executive and the Minister to be contestable in court.

Summary and Conclusions 113

All the mining acts provide for the making of special agreements between the state and the investor. This is typical of large-scale mineral development. Parliamentary ratification of these agreements is not usually necessary but they are, of course, subject to parliamentary scrutiny.

Deficiencies were noted in the mining acts of some countries, and in particular, the need to deal more adequately with the small-to-medium-size mining operations which are not artisanal but yet not large enough to accomodate the full range of economic developmental concerns, which are the focus of attention of most mining acts. It was stressed that excessive developmental demands on small operations can overburden them and render them ineffective.

Once the institutional arrangement establishing the conditions under which property rights could be aquired and protected is in place, investment in mining development by private holders of capital will only be made if, first, there is the opportunity for the project to generate a surplus, and secondly, if the split of that surplus between the State and the investor is a fair one. These will be determined by the deposit characteristics, the expected cost-price ratios and the fiscal arrangements for dividing the surplus.

In chapter 3, the various instruments used by the State to effect the division of the surplus were isolated for scrutiny. These instruments were examined against the general objectives of government policy, and an *a priori* judgement was made with respect to the likely impact of those instruments on the attainment of the set objectives.

The exhaustible, non-renewable nature of minerals is recognised as one of the principal reasons for governments attempting either to maximise revenues from their development, or to conserve their use. Government may also be concerned about immediate liquidity, or sponsoring some form of regional development, or the mine contributing to national development in an integrated way. The simultaneous attainment of all of these objectives can be quite difficult, and the government therefore should be clear on the ranking of its priorities, and should use instruments which are most effective in attaining them.

The principal instruments used by states to extract their portion of the surplus are royalty, income tax, withholding taxes, special taxes and duties. The discussion showed that royalty, as front-end levy, is the surest means of obtaining government revenues, but it contributes to operational cost, and hence immediately reduces the available profit to the project. Front-end royalties can be therefore harsh on marginal projects, and should be used with care. However, it was noted that because of this distortionary effect on the project economics, front-end royalties can be conservationist in terms of postponing the use of resource to the future, if this is believed to be in the interest of the State. In some cases it could also be depletionary by increasing cut-off grades, hence rendering a large resource base unusable because the remaining average grade may fall well below conventional or future technologically recoverable levels. All front-end charges, such as export levies, have the same effect. The point was further made that if a royalty is levied, it should be done on a value basis, thus taking account of movements in both volume of production and price. In cases where value is difficult to determine because the mineral is bound up in a vertically integrated operation, the Jamaican solution of indexing bauxite prices to the price of the final aluminium product for royalty purposes is advisable.

All income or profit-related taxes take the cost of production into account, and therefore permit the deposit characteristics and the environment to be the sole determinants of the level of the surplus. However, because several cost variables are at work, and more particularly since the investor is in control of decisions on what costs are to be reflected, the profit-related taxes can become an uncertain source of government revenue. The transfer pricing phenomenon is not to be underestimated, and requires a fair degree of sophistication on the part of the State to control. Governments must invest heavily on developing the capability of the tax collecting authorities if this problem is to be properly handled.

Profit-related taxes which are not progressive do not provide for government a large enough share of any economic rent generated on account of the exploitation of its superior deposits or in cases of an unexpected price boom. It is in this context that the Resource Rent Tax, which assumes the achievement of a certain base level rate of return by the investor before the marginal tax rate is increased, has some appeal provided that there is an efficient tax collecting machinery. The general conclusion was that the government should levy a small front-end royalty, but combine it with income and resource rent taxes.

An important element in determining the mix of instruments to be used is the tax treatment accorded to the investor in his home country. One must bear in mind the taxable limit of the company at home, to ascertain creditability. Careful guidance might be necessary on the tax treatment in the investors' countries.

The presence of foreign investment makes it essential for the State to be clear on its use of the withholding tax, whether on interest income or dividends. Inasmuch as interest income is independently determined, and free of production cost considerations, there is a school of thought which would have it taxed at the general corporate tax rate. However, if this is done one must be aware that it is to be reflected in the cost of loans to the project. As an instrument to encourage local reinvestment, a dividend withholding tax may be levied, but it must be recognised that this is simply translated by the investor into a higher expected rate of return when he makes the decision to invest. Hence, the level of the tax is an important consideration for the State. The level of withholding taxes must also bear some relationship to the investor's tax treatment at home. These points highlight the necessity for less developed countries to conclude double-taxation agreements with the investor's home country, so that fiscal packages can be put together, which will achieve for them and the investor optimal returns from the project.

Chapter 4 focused on the actual performance of the fiscal instruments in the division of the project surplus. In this context, the economic impact of both fiscal and some non-fiscal provisions was considered. Of particular importance was the evaluation of varying levels and modes of financing of the participation by the State in the mineral project. The enquiry sought to illuminate the issue of the financial implications of various expressions of sovereignty by the respective states over their natural resources. At all times, after isolating the degree of profitability which was conferred by the deposit and production characteristics, the fundamental question to be answered was what portion of the project surplus would go to the State, and what return the investor was likely to make on his investment.

The first and very obvious conclusion which emanated from the analysis was that despite identical cost-revenue assumptions for each legislative regime, the returns to the project, government and investor vary for all corresponding prices. This simply infers that the varying returns are purely due to the elements of the fiscal package, which were basically different from one state to the next. In the case where 100 per cent of the equity was owned by the foreign investor, the government take was noted to be, an average, between 55 and 70 per cent for the projects which were commercially viable.

In all the cases except Quebec, the potential government take in the marginal, non-viable projects was extraordinarily high, ranging from 61 to 100 per cent. The Quebec scheme gave the government take of 37 per cent in the marginal case, and held out the best prospect for project development, if for one reason or another, the company decreased its average expectation of its desired level of return. It was assumed that a project real rate of return of below 10 per cent could be considered marginal. The Quebec scheme permitted the project to achieve a 9.7 per cent IRR with the US$300 price assumption. Sierra Leone only gave a 2.5 per cent project IRR, while in Zambia, it was 3.7 per cent, Tanzania 6.7 per cent, Perak (Malaysia) and Botswana 6 per cent and Papua New Guinea 8 per cent.

The overriding point to note in the above result is that all the statutory regimes, except in Quebec and Papua New Guinea, are exceptionally hard on marginal projects, principally on account of large front-end levies. The point must be made that for poor countries with limited financial resources, a sound fiscal mining policy is one which countenances the putting on stream, where possible, of many small mining projects, and imposing on them as little front-end, distortionary levies as possible. Secondly, since the smaller deposits are statistically more numerous, but of little interest to the large foreign transnational, then even the post-income charges must seek to encourage development of this size. The underlying assumption, of course, is that the marginal deposit is worth developing in the national interest.

At the US$450 and US$600 price levels, if one assumes a 10 per cent real rate of return as acceptable, all the regimes are conducive to investment. In the context of the model described, this means that all the regimes are conducive to investment when the ratio of operation cost to gross revenue is in the vicinity of 55 per cent and less. However, note that in the case of Sierra Leone, the 45 per cent dividend withholding tax *may* reduce the return to the investor considerably from the project rate of return.

The capital allowances used by the various states varied considerably, and it was demonstrated that they significantly affected the year-to-year flows. The immediate write-off in Zambia and Malaysia produced lower government income tax receipts in the earlier years than in later years, while for regimes where capital allowances are distributed evenly throughout the life of the project, the government tax receipts were more steady. The stage of development of the country, the nature of the fiscal impositions, size of project, financing requirements and the attractiveness of the country to investment are central to the decision, on the part of the State, as to the appropriate level of capital allowances which it should grant the investor.

The results of the analysis showed great variations in the government revenue-mix of the various countries. Using US$450 gold price as an example for the 100 per cent foreign investor participation, it was noted that Quebec will get over 40 per cent of its take from the profit-based mining tax, while Papua New Guinea will only get 0.4 per cent of total government take from royalties. For Zambia, the mineral tax is 30 per cent of total government revenue from the project, while the royalty gave Botswana 13 per cent, Sierra Leone 12 per cent, Perak (Malaysia) 10 per cent and Tanzania 5.2 per cent of the total government receipts. The implication is that during bad years, the Papua New Guinea and Quebec profit-based schemes will give low government revenues from royalties, but higher revenues during boom times. For the other regimes, the flows move more or less proportionately with varying price levels.

In a number of countries, it is *en vogue* to view government participation in projects as a *sine qua non* for revenue maximisation. This phenomenon was investigated, and apart from qualification consideration which addresses the question of the manipulation by the foreign investor of the economic and financial indices of performance, hence obfuscating their true liability for tax, the need of the State to dictate the pace, level, form and integration of the project in the general economy was seen as a powerful reason for the State to participate in mineral development.

In the case of the regime which allowed for 51 per cent fully paid government equity, it was to be noted that the return on government equity was essentially the return to project equity. The implication is, of course, that among other things, any government wishing to buy equity in a mining venture, must compare the rate of return on its equity with those of other investments. Cognisance must be taken of the fact that generally, as the project grew more profitable, the share of the government take rose as a result of participation, but in none of the cases, did that 'take' increase by more than 25 per cent for the government having acquired 51 per cent of the equity, compared to a situation in which it did not participate. The main merit of government equity participation would appear to be to obtain effective control rather than maximise revenue.

Of course, it must, however, be admitted that the theoretical revenue flows in the absence of this control may not be realisable in their entirety, especially in the context of an ill-trained, insufficiently equipped, and disorganised government bureaucracy. The obvious remedy, which may not make as many demands on scarce resources, is in the development of a professional cadre which could monitor the activities of the enterprise. In some circumstances this service can be bought, but when it comes from external sources, it must be accompanied by the vigorous training of counterpart local staff.

Mining rights have a value, not only on account of the previous work which may have been done to establish the prospect, but also from the standpoint of the potential mineral deposit, and the exclusion of others from having access to that deposit. It is therefore not uncommon for less developed countries to seek some recognition for the value of the rights which they are conferring. This may take the form of bonuses at the time of signing or by the establishing of a carried interest in a future commercial operation. The carried interest may involve the acquisition by the State of a share interest in the capital stock of the company at a discount or gratis, within certain time frames. This equity received in consideration for mining rights was referred to as 'free' equity in the first section of chapter 4, and the analysis indicated that in the case where 51 per cent government participation is required, the acquisition of 20 per cent of this equity through the 'free' route allows, on average, a 2 to 3 per cent increase in the government take in the regimes considered. 'Free' equity is a useful means by which the government can get in on the ground floor and increase its information base.

When loan financing was considered, the analysis revealed that the introduction of debt produced higher project real IRR's for all regimes because the interest rate was a tax deductible item. The results further showed that there was, on average, a 2 per cent decrease in the government take when the government 31 per cent carried interest was financed out of future dividends as opposed to when it was paid for directly by the State. This decrease was occasioned principally on account of the fact that, in the model used, the rate of interest (11 per cent) on the outstanding unpaid equity was above the inflation rate (7 per cent). (Alternative assumptions of inflation/dividends rates could well have given different results but the intention was to compare equity versus loan financing.)

The obvious attraction of this form of financing of government equity is that it makes fewer demands on the State to tie up scarce capital resources, when they could be deployed elsewhere. However, in strict economic terms, government funds diverted from the mining project should at least not only earn the project internal rate of return, but also the interest paid by the government on borrowed funds for its equity. In some sense, the project IRR plus the interest paid by the government on loans to purchase equity reflect in a sense the opportunity cost of government funds *vis-à-vis* the mining project. Of course, in situations where the financial resources are just not available locally, this mechanism of financing government participation is a very attractive one, especially in circumstances of a long projected mining life for the deposit. The State can therefore acquire major participation, at little cost to itself (foregone dividend only) over a time frame which still permits it to have benefited from the dynamic economies of learning-by-doing, so that it could be in a better position to convert its nominal ownership into real control.

It is absolutely essential though, for one point to be re-emphasised, and that is – although the form of financing can change the economics of the project, the ultimate strength of the project lies in its ability to generate economic rent. After all, the method of financing merely redistributes the risk, it does not eliminate it.

Mineral development is a high-risk business, which not only involves the uncertainties surrounding the exploration, definition, mining and processing of an ore deposit, but also the economic and political risks which affect normal business activity, including large fluctuations in the price of final output. While the financing method can redistributes the risks, the investor seeks other assurances to eliminate or reduce the level of his risk. Ignorance of the environment, in terms of the operations of its laws, the stability of its institutions, is perhaps the greatest risk factor which must be accomodated in the investor's decision to commit his capital. The long periods over which the investment may be committed before returns can be had, magnify his perceptions of the risks, imagined or real.

Amid these circumstances, the foreign investor, in contemporary times, views participation with the host government as desirable for the conduct of a safe business in the developing country. Governments, apart from wanting to exercise greater sovereignty over the mineral resources, have viewed partnership with the foreign investor as being useful, especially in terms of gaining access to superior technology and accompanying skills. As a consequence, the joint-venture arrangement is the principal way in which state-investor relationships have developed in the mineral sector. Generally, the State does not invest in exploration, but reserves the right to participate in a joint-venture company, in a major or minor way, after an ore deposit has been established.

A related issue is the control and management of the joint enterprise. At best, the State seeks the situation where, even with a minority shareholding, it has control over the Board of Directors, though generally, consensus is normally achieved with board representation reflecting shareholding, but with minority safeguards on specific crucial policy issues. The investor is accorded the executive right to manage the company up to a certain point in time, depending on the size of project, amount of investment, level of reserves, and the mineral commodity in question. These are accompanied by a responsibility on his part to localise the industry, by training and employing nationals of the host state at all levels, and by promoting and encouraging the use of local materials.

The over-all objective is to secure for the investor an acceptable return on his investment, while compensating the government for the use of its exhaustible stock of mineral resources. The mining development agreement provides for fiscal levies which are intended to be in place during the life of the project. These fiscal levies form part of the government's share of the economic rent. Basically, the investor seeks rapid recoupment of his investment and the freedom to repatriate the capital and the earnings arising therefrom. The State also expects the investor to reinvest, locally, the earnings he makes on his investment. As a consequence, foreign exchange accomodation outside of the applicable law and freedom from withholding taxes are part of the conditions which a foreign investor may present to a host state, which in turn views them as matters to be surrendered only grudgingly.

There are no hard and fast rules with regard to the choice and level of application of any particular instrument. The overriding consideration is that the package must satisfy both parties. It is obvious that any contract which confers unequal advantage to one side runs the great risk of being unstable over time. Equity in treatment is equally important as equity in holdings!

The large vertically-integrated conglomerate is the dominant agent of mineral development, depending on the particular mineral in question. By and large, the processing and marketing of minerals are concentrated in the developed countries. The attraction of the significant value – added at the processing stage, and its ability to promote industrial development, has caused under-developed countries to increase their demands for more local processing of their primary mineral resources. This gets reflected in many mining development agreements, with the State granting concessions to the company to vertically integrate locally. The merits of domestic processing of raw materials will ultimately depend on the circumstances of each country hinging on a comparison of the social costs and benefits.

An undesirable feature of the operations of the transnational corporation, as far as the host government is concerned, is the practice by those companies of freely transferring prices and cost centres so that they can globally maximise profits. Mining agreements therefore normally reflect the State's concern for arresting this tendency by making stipulations for corporate disclosure and arms-length trading practices. Ultimately, the State must develop a technical capability to monitor these companies, if transfer-pricing is to be eliminated. This capability can be bolstered if local practitioners are initially involved in the workings of the enterprise. This need to develop a knowledge of the conduct of business is perhaps one of the more important reasons for states to insist on localisation requirements. Sovereignty, expressed as the developing of a national capability to manipulate the instruments of the productive process for economic gain, presupposes that the State must be involved, not only as a regulator, but as an active participant in the economic process.

The negotiation of mineral development agreements can be tedious. The conflicting objectives of the participants – the State and the investor – have to be settled in an atmosphere of incomplete knowledge. The deposit characteristics, and the sociopolitical and economic conditions of the country during the exploitation period, are all unknown at the time of the negotiations. It is only with the explicit treatment of the number of variables which are built into the financial model, and their realistic accomodation, that any lasting agreement would be made. This calls for the anticipation of changing circumstances and the way in which the two parties are prepared to react to them. Given that all the possible circumstances cannot be predicted, the agreements must admit of some flexibility and make allowance for review and changes.

The agreement must also take cognisance of the reality of what may be considered irreconcilable differences at the level of the local procedures and courts. The insertion of a clause which gives the parties access to a judicial process according to international law and practice, is a strong incentive for the investor to reduce his perception of the risk which may surround his investment. A reduced-risk estimation translates itself into lower levels of expected returns, thus making it possible for more potential targets to be explored. A reduced perception of risk also makes the atmosphere of the negotiations more amiable, and hence more conducive to settlement.

The government in the under-developed country should view the negotiation process as an input in economic development. The government should select and maintain a group of professionals who are competent in their field, and can gain the respect of the opposite party, since respect is an essential key in the case in which a negotiation can take place. The essential support which the government negotiation team needs is the presence of a fairly comprehensive set of laws which define how, when and where an investment in the mineral sector can be undertaken in the State. The State further needs to be clear about its objectives, so that coherence can be maintained in its posits.

Sovereignty over natural resources as conceived in the UN declaration is not simply a term which refers to the right to ownership of the natural resources within specified boundaries, which define a state. It is a term which refers alia to the State's ability to organise and manipulate the co-operant resources, which would allow the transformation of a mineral endowment to usable products. It is a term which implies the maximum economic benefit that can be derived from transforming the resources.

In the context of mineral development in the under-developed country, where the capital and the end-users of the final product are of an external origin, the expression of sovereignty takes on a very complex and delicate nature. Apart from the ability to dispose of mineral rights in a form consistent with national policy, whether it is viewed in terms of economic returns or locational developmental or security prerogatives, the developing country is faced with the dilemma of having to contemplate the subrogation of its laws to international law and practice, because the foreign investor needs safeguards, and in some peculiar circumstances, to the workings of the global political game. It is in this context that the posit must be made that the expressions of sovereignty over natural resources cannot be expressed in terms of a specific formula. Sovereignty can only be expressed in the peculiar circumstances of each individual country. However, the common thread that runs through the concerns of all under-developed states, is the need to maximise the economic benefits from the exploitation of their mineral resource, and it is to this concern that this work has been largely addressed. These countries will have to maintain a capability in identifying and seeking the conditions which would permit investment, while safeguarding for the State, the maximum benefits from the use of its exhaustible stock of mineral resources.

Appendices

Appendix 1	Financial Provisions and Charges	124
Appendix 2	Requirements in Application for a Mining License	134
Appendix 3	Selected Computer Print-outs of Results	136
Appendix 4	Depreciation Methods	164
Appendix 5	Numerical Example of a Cash Flow Based Rent Resource Tax	167
Appendix 6	Explanation of Some Basic Financial Concepts	171

Appendix 1

FINANCIAL PROVISIONS AND CHARGES

(A) SIERRA LEONE

Fees:	Prospecting right	= Leones 15	
	Alluvial gold mining licence	= Leones 4	
	Gold trader licence	= Leones 10	
Rents:	Exclusive propsecting	= Leones 20	per sq. mile per annum
	Mining right (river locations)	= Leones 70.4	per mile per annum
	Mining lease (various classes but maximum of)	= Leones 2	per acre for precious metals and lodes

Royalty

	Chromite	Ilmenite	Iron
Fixed charges per unit	Le 0.08	Le 0.10	Le 0.06

Charge per unit based on price

Columbite (ores containing 65 units of combined Cb_2O_5, Ta_2O_5)

When price of Cb_2O_5	Code	Incremental royalty	Total royalty
≤ Le 10	(p)	0.04p	.04p
> Le 10 ≤ Le 20	(p1)	0.12p1	.04p + .12p1
> Le 20 ≤ Le 40	(p2)	0.16p2	.04p + .12p1 + .16p2
> Le 10 ≤ Le 60	(p3)	0.30p3	.04p + .12p1 + 16p2 + 13p3
> Le 60		0.5p4	.04 + .12p1 + .16p2 + 3p3 + 5p5

Charge per unit based on production

Rutile

For production	Charge	
< 30,000 tons per year	Le 3	per tonne
30,000t ≤ 60,000t	Le 5	per tonne
> 60,000 tons per year	Le 6	per tonne

125

Charge per unit based on value:	Gold, Platinum and Diamond 5% ad valorem	
Income tax:	45%	
Withholding tax:	45% on dividends and interest	
Capital allowances:	Greater of 5% or $\left(\dfrac{\text{output}}{\text{reserves}}\right)$	
Surtax:	15% of income tax	
Diamond industry profits tax:	27½% of profits	
Iron ore concession tax:	5% of chargeable income	

(B) TANZANIA

Fees:	Prospecting licence (PL)	Shs 40	Renewal of PL Shs 20
	Max. fee of any kind	Shs 100	
Rents:	Exclusive prospecting licence		per sq. ml
	Claim	Shs 100	
	Lease	Shs 10	
		Yr 1	Yr2 Yr3 Yr4
	per acre	Shs 1	Shs 1.50 Shs 2.00 Shs 2.50
Royalty:	Diamond, other precious and semi-precious stones	15% ad valorem	
	Base metals, platinum and mica	5% ad valorem	
	Gold and silver	1½% ad valorem	
	Kaolin	2½% ad valorem	
	(Value of assessment is price−cost of handling, transport, insurance to point of sale)		
Taxes:	General company rate	= 50%	
	Foreign company rate	= 55%	
	Mining company rate	= 22½% year 1 to 4	
	For specified minerals (Cu, Au, Sn, W, $CaCO_3$, coal, vermiculite, bentonite, mica, magnesite).		
	Mining company rate	= 45% per year 1 to 4	
		= 50% after four years	

Withholding tax:	*Dividends*	*Interest*	*Management fees*	*Royalty on patents etc.*
	10%	12½%	20%	20%
Capital allowances:	Including write-off of exploration 40:10:10:10:10:10:10
	For 'specified minerals' deduction allowed as expenditure is incurred
Sales taxes:	For precious stones sold locally: 24%
	This is remitted if stones are exported
Diamond levy:	Levy = 5% of gross value, if gross production is greater than 10 m shillings

(c) ZAMBIA

Fees : Prospecting licence (Industrial Minerals) K6 (US$3.85) per sq. km
Mineral permit (Industrial Minerals) K25 (US$16.02) per hectare

Royalty : There are no royalties

Tax : 45% on assessable income less mineral tax

Withholding tax : 20% on dividends and interest

Capital allowances: For old mines

Plant and machinery	Industrial buildings	Low cost housing
30% on declining balance	10% initial 5% annual	10% initial 10% annual

For new mines (after 1953)

Capital expenditure allowed in full in year incurred

Mineral	*Cu*	*PbZn*	*Amethyst/beryl*	*Au, Bi, Se, Co, Hg, Cd*
Mineral tax				
Rate %	51%	20%	15%	10%

If return on equity is less than 12%, the mineral tax is refunded to achieve this target.

(D) BOTSWANA

Rents:	For mining claim other than for precious stones				
Per month charge:	Yr1	Yr2	Yr3	Yr4	Yr5
	£1.25	£2.5	£3.8	£5.0	£6.3
	For precious stones mining claim:			£63 per month	
	For alluvial precious stones claim:			£31.5 per month	
Royalty:	Precious stones			10%	
	Precious (base) metals and semi-precious stones			5%	
	Industrial minerals and oil shale			3%	
Tax:	Company rate			35%	
Withholding tax:	On dividends and interest			15%	
Capital allowances:	Written off over life of mine (which should not exceed 30 years)				

(E) PAPUA NEW GUINEA

Rent: Special mining lease K2.50 per hectare per annum

Royalty: 1.25% f.o.b. value

Tax:

(A) *Small-scale mining:* (mining on gold mining leases or other mining leases)
   For resident company 33⅓%
   For non-resident company 45%

   Capital expenditure allowance for small-scale mines (max. annual allowance = 5%)

   $$\frac{\text{residual capital expenditure (RCE)}}{\text{remaining life of mine}} \quad \text{or} \quad \frac{\text{RCE}}{20} \quad \text{whichever is less}$$

   Note: Plant and other articles excluded from definition of RCE, and are allowed for by various schedules

(B) *Large-scale mining:* (mining accomplished under a special mining lease)
   For resident company 33⅓%
   For non-resident Company 45%
   Note: statutes have 35 and 48 respectively, but the lower rates currently apply

   Capital expenditure allowance

   Initial allowance: maximum of 25%

   Annual allowance: $\frac{\text{Residual expenditure (RE)}}{\text{remaining life of mine}}$ or $\frac{RE}{10}$ whichever is less
   Therefore maximum of 10% for capital expenditure

   Annual allowance (for previous exploration) $= \frac{\text{residual exploration expenditure (REE)}}{\text{remaining life of mine}}$ or $\frac{REE}{5}$

Therefore maximum of 20% for exploration expenditure

Additional profits tax (APT): $(70-n)$ NCR, where, n = current normal corporate income tax rate
NCR = accumulated value of net cash receipts

Accumulation Rate (that is, the threshold rate of return) is either
20% or prime rate of interest in USA averaged over year of income plus 12%
$$NCR = Y - D - C - Ex - P - T - \Delta I$$
Note Y = income in current year from sales, reduction in stocks or sale of assets
D = allowable deductions apart from depreciation, interest etc.
C = allowable capital expenditure made in year
Ex = Exploration expenditure for current period and for last 11 years
P = Any equipment bought in year of income
T = Normal corporate income tax paid
ΔI = Average increase in inventory of spares, mining products and consumable stocks

To accommodate expenditure incurred prior to production and which must bear interest (R)
$NCR = A(100\% + R) + B$, where A = net assessable cash receitps in all periods; and
B = net assessable cash receipts in year of income

These are then adjusted for foreign exchange changes by multiplying by $\frac{F}{E}$,

where H = average of buying and selling rates for the two currencies in the year of income
E = average of buying and selling rate for the two currencies in the previous year

$$NCR = \frac{F}{E}(A(100\% + R) + B$$

APT payable = $70 - n$(NCR)

Withholding tax: 15% on dividends
45% in interest to non-residents

(F) MALAYSIA

Fees:	Miscellaneous maximum	M$ 30.00
Licence charges:	Application for a prospecting licence or permit or mining lease or renewal of lease	M$ 20.00
Rent:	On mining lease or mining certificate	M$ 4.00 per acre
Premiums:	(1) *New alienations*	
	(a) Lease not exceeding 10 years	M$ 100.00 per acre plus $10 per acre for each year of the term in excess of 10 years
	(2) *Renewal of mining leases*	
	(a) For each year of renewal	M$ 10.00 per acre
	(b) Penalty premiums for late applications	

Months	Late	6 months	6 to 9 months	Over 9 months
Premiums per acre		M$ 10	M$ 15	M$ 20

(c) For bad labour record M$ 10 per acre

(3) *Prospecting permits and prospecting licences*

(a) Prospecting licence: Every 1000 acres and under M$ 250.00

(b) Prospecting permit:

Size:	Less than 100 acres	> 100 acres < 500 acres	> 500 acres < 1000 acres	1000 acres
Premium:	M$ 100	M$ 50	M$ 150	M$ 200

Royalty:	Gold	5% ad valorem
Tin duties:	An export tax is charged on the export (delivery to smelter) of all tin concentrates when the price of tin metal exceeds M$ 400 per picul according to the following formulas:	
	On the first M$ 400 per picul	ad valorem NIL (January 81)
	Plus on the next M$ 50 per picul	ad valorem 20%
	Plus on the next M$ 50 per picul	ad valorem 25%
	Plus on the next M$ 50 per picul	ad valorem 30%
	Plus on the next M$ 50 per picul	ad valorem 35%
	Plus on the next M$ 50 per picul	ad valorem 40%
	Plus on the next M$ 50 per picul	ad valorem 45%
	Plus on the balance	ad valorem 50%

The tax according to the above formula is charged on the export of tin concentrates and therefore to arrive at the quantum of tax per unit of contained tin the figures arrived at by applying the formula have to be increased by approximately 32.5% in respect of concentrates from underground mines which contain approximately 70% tin

Tin profits tax: Tin profits tax payable on a sliding scale, i.e.
5% on the first M$200 000 of taxable tin profits
10% on the next M$200 000 of taxable tin profits
15% on any taxable tin profits exceeding M$400 000

Income tax: 40% of taxable profits

Development tax: 5% on taxable profits

Withholding tax: 15% on dividends, management and technical services
(Since 1978 interest has been freed of withholding tax)

Capital allowances and investment incentives

(1) For the period 1978–86, all industries were given an 80% accelerated allowance on plant machinery apart from the 20% initial allowance already in place

(2) For equity restructuring where at least 30% of the equity of the enterprise is held by bumiputra (local Malays), 40% for non-bumiputra Malaysians, and 30% for foreigners, the companies can gain a 5% points reduction from the 40% company tax

(G) QUEBEC (CANADA)

Fees:	Prospecting licence	$10.00
	Development licence and renewal	$10.00
Rent:	Under development licence	$0.25 per acre for 1st and 2nd years
		$0.75 per acre for subsequent years
		$1.00 per acre for late renewal
		$1.50 per acre for further late renewal
	Under mining leases:	$1.00 per acre
		If work does not start in two years, the following rental rates shall apply:
		Yr 3 4 5 6 7 8 9 10
		$ 2 2 2 3 3 4 4 5 6
	Under exploration permit:	Variable, but at least $1.50 per sq. mile
Quebec mining tax:	Income for mining tax	Marginal rate
	Up to $250 000	Exempt
	$250 000 – $3 250 000	15%
	$3 250 000 – $10 250 000	20%
	$10 250 000 – $20 250 000	25%
	Over $20 250 000	30%
Quebec corporate tax:	12% of income for Quebec corporate income tax	
Federal corporate tax:	36% of income for Federal corporate income tax	
Processing allowance:	8% of original cost of processing assets if ore is concentrate; and 15% of original cost if concentrate is further processed in Quebec. Allowance must be greater than 15% but less than 65% of income for processing allowance	

Cash Flow Format for Calculating Taxes

Federal corporate income tax
Revenue
− *Operating cost*
Net income before allowances
− *Capital cost allowance*
Resource profits
− *Resource allowances*
Interest payments
− *Exploration expenditures*
− *Development expenditures*
− *Income for earned depletion*
− *Earned depletion allowance*
Income for federal corporate income tax

Quebec corporate income tax
Revenue
− *Operating cost*
Net income before allowances
− *Depreciation allowances*
Resource profits
− *Resource allowances*
Interest payments
− *Exploration expenditures*
− *Income for earned depletion*
− *Earned depletion allowance*
Income for Quebec corporate income tax
Quebec corporate income tax

Quebec mining tax
Revenue
− *Operating cost*
Net income before allowances
− *Depreciation allowances*
− *Exploration expenditures*
− *Income for earned depletion*
− *Earned depletion allowance*
− *Income for processing allowance*
− *Processing allowance*
Income for Quebec mining tax
Quebec mining tax

Federal capital cost allowance: Provincial depreciation allowance = 100% of capital expenditure incurred on mine, mill and infrastructure after start of commercial production for new mines

Resource allowance: 25% of resource profits
Development expenditures: 30% per year on a declining balance basis for mine access capital cost, incurred before the start of commercial production
Earned depletion allowance: $1.00 for every $3.00 of expenditure on exploration, mine access, plant and machinery during the development state, or associated with a major expansion of plant, machinery and mill facility; and processing machinery and sustaining capital costs. Maximum of this allowance is 25% of income for earned depletion in any year

Appendix 2

(Extract from Art. 32 of the Mining Act of Botswana (1976)

REQUIREMENTS IN APPLICATION FOR A MINING LICENCE

Applicant must give:

(a) his full name and nationality, or, in the case of an application by a partnership or other association of persons, the full names and nationalities of all partners or of all such persons, or, in the case of an application by a corporate body, the registered name of such body;
(b) in the case of a corporate body, the full names and nationalities of the directors and the full name and nationality of any shareholder who is the beneficial owner of 5 per cent or more of the issued capital;
(c) full information as to his financial status, technical competence and experience;
(d) the number of his prospecting licence;
(e) the name of the mineral which it is intended to mine;
(f) details of the mineral deposit and a comprehensive report thereon which shall include details of all known minerals, proved, estimated and inferred ore reserves and mining conditions;
(g) details, illustrated by an approved plan, of the area in respect of which the application is made;
(h) the period for which the lease is sought;
(i) a technological report on mining and treatment possibilities and the intention of the applicant in relation thereto;
(j) a proposed programme of mining operations which shall include:

  (i) the date by which the applicant intends to work for profit;
  (ii) the capacity of production and scale of operations;
  (iii) the estimated over-all recovery of ore and mineral products;
  (iv) the nature of the product;
  (v) the marketing arrangements made for the sale of the mineral product; and
  (vi) a detailed programme for the progressive reclamation and rehabilitation of lands disturbed by mining and for the minimisation of the effects of such mining on adjoining land and water areas;

(k) a detailed forecast of capital investment, operating cost and sales revenues and the anticipated type and source of financing;
(l) a programme for the employment and training of Botswana citizens;
(m) a report of the goods and services required for the mining operations which can be obtained within Botswana and the applicant's intention in relation thereto;
(n) details of expected infrastructure requirements; and
(o) such further information as the Minister may require or as may be prescribed.

Appendix 3

SELECTED COMPUTER PRINT-OUTS AND TABLES OF RESULTS

TABLE 3.1 Zambia: 31% of government equity fully paid and 20 is 'free' – gold price at US$450 on oz

	1983	1984	1985	1986	1987	1988	1989	1990	1991	1992
OUTPUT				60000-0	60000-0	60000-0	60000-0	60000-0	60000-0	60000-0
PRICE	450.00	481.50	515.20	551.27	589.86	631.15	675.33	722.60	773.18	827.31
REVENUE				33.08	35.39	37.87	40.52	43.36	46.39	49.64
ACCUMULATED EXPL EXP	5.00									
DEVELOPMENT COSTS	5.00	16.05	11.45							
OPERATING COSTS				14.70	15.73	16.83	18.01	19.27	20.62	22.06
MINERAL RENT				3.31	3.54	3.79	4.05	4.34	4.64	4.96
CAPITAL ALLOWANCE				37.50						
VENTURE DEBT REPAYMENT										
LOAN	7.00	11.23	8.01							
INTEREST	0.77	2.01	2.89	2.41	1.92	1.44	0.96	0.48		
LOAN REPAYMENT				4.37	4.37	4.37	4.37	4.37	4.37	
INT WITHHOLDING TAX	0.15	0.40	0.58	0.48	0.38	0.29	0.19	0.10		
TAXABLE INCOME	−0.77	−2.78	−5.66	−30.50	−16.30	−0.49	17.00	19.27	21.13	22.61
INCOME TAX							7.65	8.67	9.51	10.18
JT VENTURE DIVD				8.29	9.82	11.43	5.47	6.22	7.25	12.44
INVESTOR DIVIDENDS				4.06	4.81	5.60	2.68	3.05	3.55	6.09
GOVT DIVIDENDS				4.23	5.01	5.83	2.79	3.17	3.70	6.34

INVESTOR CASH FLOWS										
INVES NCF	−3.38	−5.80	−4.85	3.25	3.85	4.48	2.14	2.44	2.84	4.88
INVES REAL NCF	−3.38	−5.42	−4.24	2.65	2.94	3.20	1.43	1.52	1.65	2.65
DISD 5% INVES REAL NCF	−3.30	−5.04	−3.75	2.24	2.36	2.44	1.04	1.05	1.09	1.67
INVES REAL NCF IRR	14.35	14.35	14.35	14.35	14.35	14.35	14.35	14.35	14.35	14.35
GOVT CASH FLOWS										
GOVT NCF	−0.24	−0.62	−0.90	8.83	9.90	11.03	15.22	16.89	18.56	22.70
REAL GOVT NCF	−0.24	−0.58	−0.78	7.21	7.55	7.86	10.14	10.52	10.80	12.35
DIS 5% GOVT REAL NCF	−0.23	−0.54	−0.69	6.07	6.06	6.01	7.39	7.29	7.13	7.77
GOVT SHARE RNCF				73.10	71.99	71.10	87.65	87.38	86.72	82.32
GOVT CARRD INT FLOWS	−1.17	−2.11	−1.96	2.57	3.05	3.54	1.70	1.93	2.25	3.86
REAL GOVT CARRD INT FLS	−1.17	−1.98	−1.71	2.10	2.32	2.53	1.13	1.20	1.31	2.10
GOVT CARRD INT IRR	30.77	30.77	30.77	30.77	30.77	30.77	30.77	30.77	30.77	30.77
REAL TOTAL SURPLUS	−3.62	−6.00	−5.02	9.86	10.49	11.06	11.57	12.04	12.45	15.00
REAL PROJECT SURPLUS	−3.77	−6.37	−5.52	6.76	7.49	8.15	3.65	3.88	4.22	6.76
PROJECT REAL IRR	30.77	30.77	30.77	30.77	30.77	30.77	30.77	30.77	30.77	30.77

	1993	1994	1995	1996	1997	1998	1999	2000	TOTAL
OUTPUT	60000*0	60000*0	60000*0	60000*0	60000*0	60000*0	60000*0	60000*0	899999*0
PRICE	885.22	947.18	1013.49	1084.43	1160.34	1241.56	1328.47	1421.47	
REVENUE	53.11	56.83	60.81	65.07	69.62	74.49	79.71	85.29	831.17
ACCUMULATED EXPL EXP									5.00
DEVELOPMENT COSTS									32.50
OPERATING COSTS	23.61	25.26	27.03	28.92	30.94	33.11	35.43	37.91	369.41
MINERAL RENT	5.31	5.68	6.08	6.51	6.96	7.45	7.97	8.53	83.12
CAPITAL ALLOWANCE									37.50
VENTURE DEBT REPAYMENT									
LOAN									26.25
INTEREST									12.88
LOAN REPAYMENT									26.25
INT WITHHOLDING TAX									2.58
TAXABLE INCOME	24.20	25.89	27.70	29.64	31.72	33.94	36.31	38.85	
INCOME TAX	10.89	11.65	12.47	13.34	14.27	15.27	16.34	10.48	147.72
JT VENTURE DIVD	13.31	14.24	15.24	16.30	17.44	18.66	19.97	20.37	197.46
INVESTOR DIVIDENDS	6.52	6.98	7.47	7.99	8.55	9.15	9.79	10.47	96.75
GOVT DIVIDENDS	6.79	7.26	7.77	8.31	8.90	9.52	10.19	10.90	100.70
INVESTOR CASH FLOWS									
INVES NCF	5.22	5.58	5.97	6.39	6.84	7.32	7.83	8.38	63.38
INVES REAL NCF	2.65	2.65	2.65	2.65	2.65	2.65	2.65	2.65	24.22
DISD5% INVES REAL NCF	1.59	1.51	1.44	1.37	1.31	1.24	1.19	1.13	10.59
INVES REAL NCF IRR	14.35	14.35	14.35	14.35	14.35	14.35	14.35	14.35	

GOVT CASH FLOWS									
GOVT NCF	24.29	25.99	27.81	29.76	31.84	34.07	36.45	39.01	350.58
REAL GOVT NCF	12.35	12.35	12.35	12.35	12.35	12.35	12.35	12.35	163.61
DISS% GOVT REAL NCF	7.40	7.05	6.71	6.39	6.09	5.80	5.52	5.26	96.47
GOVT SHARE RNCF	82.32	82.32	82.32	82.32	82.32	82.32	82.32	82.32	
GOVT CARRD INT FLOWS	4.13	4.41	4.72	5.05	5.41	5.79	6.19	6.62	55.97
REAL GOVT CARRD INT FLS	2.10	2.10	2.10	2.10	2.10	2.10	2.10	2.10	24.60
GOVT CARRD INT IRR	30.77	30.77	30.77	30.77	30.77	30.77	30.77	30.77	
REAL TOTAL SURPLUS	15.00	15.00	15.00	15.00	15.00	15.00	15.00	15.00	187.83
REAL PROJECT SURPLUS	6.76	6.76	6.76	6.76	6.76	6.76	6.76	6.76	79.37
PROJECT REAL IRR	30.77	30.77	30.77	30.77	30.77	30.77	30.77	30.77	

TABLE 3.2 Tanzania: 31% government equity paid out of dividends, 20% 'free' equity – gold price at US$450 an oz

	1983	1984	1985	1986	1987	1988	1989	1990	1991	1992
OUTPUT				60000■0	60000■0	60000■0	60000■0	60000■0	60000■0	60000■0
PRICE	450.00	481.50	515.20	551.27	589.86	631.15	675.33	722.60	773.18	827.31
REVENUE				33.08	35.39	37.87	40.52	43.36	46.39	49.64
ACCUMULATED EXPL EXP	5.00									
DEVELOPMENT COSTS	5.00	16.05	11.45							
OPERATING COSTS				14.70	15.73	16.83	18.01	19.27	20.62	22.06
ROYALTIES				0.50	0.53	0.57	0.61	0.65	0.70	0.74
CAPITAL ALLOWANCE				15.00	3.75	3.75	3.75	3.75	3.75	3.75
VENTURE DEBT REPAYMENT										
LOAN	7.00	11.23	8.01	2.41	1.92	1.44	0.96	0.48		
INTEREST	0.77	2.01	2.89	4.37	4.37	4.37	4.37	4.37	4.37	
LOAN REPAYMENT		0.25	0.36	0.30	0.24	0.18	0.12	0.06		
INT WITHHOLDING TAX	0.10									
TAXABLE INCOME	-0.77	-2.78	-5.66	-5.19	8.27	15.28	17.19	19.21	21.33	23.08
INCOME TAX					3.72	6.87	7.74	9.60	10.66	11.54
JT VENTURE DIVD				11.10	9.11	7.78	8.83	8.98	10.04	15.29
INVESTOR DIVIDENDS				5.44	4.46	3.81	4.33	4.40	4.92	7.49
GOVT DIVIDENDS				5.66	4.65	3.97	4.50	4.58	5.12	7.80
GOVT BORROWING	0.93	1.49	1.06							
GOVT REPAYMENT				2.83	1.62					
INVESTOR CASH FLOWS										
INVES NCF	-3.38	-5.80	-4.85	7.72	5.64	3.43	3.89	3.96	4.43	6.74
INVES REAL NCF	-3.38	-5.42	-4.24	6.31	4.30	2.45	2.59	2.47	2.58	3.67
DISD5% INVES REAL NCF	-3.30	-5.04	-3.75	5.32	3.46	1.87	1.89	1.71	1,702.31	
INVES REAL NCF IRR	22.31	22.31	22.31	22.31	22.31	22.31	22.31	22.31	22.31	22.31

GOVT CASH FLOWS										
GOVT NCF	−0.30	−0.77	−1.11	4.17	7.96	11.97	13.40	15.33	16.97	20.83
REAL GOVT NCF	−0.30	−0.72	−0.97	3.40	6.07	8.53	8.93	9.55	9.88	11.33
DIS 5% GOVT REAL NCF	−0.29	−0.67	−0.86	2.87	4.88	6.53	6.50	6.62	6.52	7.13
GOVT SHARE RNCF				35.06	58.52	77.73	77.48	79.48	79.31	75.55
TOTAL SURPLUS	−3.67	−6.57	−5.96	11.90	13.60	15.40	17.29	19.29	21.40	27.58
REAL TOTAL SURPLUS	−3.67	−6.14	−5.21	9.71	10.38	10.98	11.52	12.01	12.45	15.00
REAL PROJECT SURPLUS	−3.77	−6.37	−5.52	9.06	6.95	5.55	5.88	5.59	5.84	8.32
PROJECT REAL IRR	33.73	33.73	33.73	33.73	33.73	33.73	33.73	33.73	33.73	33.73

	1993	1994	1995	1996	1997	1998	1999	2000	TOTAL
OUTPUT	60000∞0	60000∞0	60000∞0	60000∞	60000∞0	60000∞0	60000∞0	60000∞0	89999∞1
PRICE	885.22	947.18	1013.49	1084.43	1160.34	1241.56	1328.47	1421.47	
REVENUE	53.11	56.83	60.81	65.07	69.62	74.49	79.71	85.29	831.17
ACCUMULATED EXPL EXP									5.00
DEVELOPMENT COSTS									32.50
OPERATING COSTS	23.61	25.26	27.03	28.92	30.94	33.11	35.43	37.91	369.41
ROYALTIES	0.80	0.85	0.91	098	1.04	1.12	1.20	1.28	12.47
CAPITAL ALLOWANCE									37.50
VENTURE DEBT REPAYMENT									
LOAN									26.25
INTEREST									12.88
LOAN REPAYMENT									26.25
INT WITHHOLDING TAX									1.61
TAXABLE INCOME	28.71	30.72	32.87	35.17	37.63	40.27	43.09	46.10	197.42
INCOME TAX	14.36	15.36	16.44	17.59	18.82	20.13	21.54	23.05	218.41
JT VENTURE DIVD	14.36	15.36	16.44	17.59	18.82	20.13	21.54	23.05	107.02
INVESTOR DIVIDENDS	7.03	7.53	8.05	8.62	9.22	9.87	10.56	11.30	111.39
GOVT DIVIDENDS	7.32	7.83	8.38	8.97	9.60	10.27	10.99	11.76	3.49
GOVT BORROWING									
GOVT REPAYMENT									4.45

INVESTOR CASH FLOWS									
INVES NCF	6.33	6.77	7.25	7.76	8.30	8.88	9.50	10.17	86.75
INVES REAL NCF	3.22	3.22	3.22	3.22	3.22	3.22	3.22	3.22	37.07
DISD5% INVES REAL NCF	1.93	1.84	1.75	1.67	1.59	1.51	1.44	1.37	19.25
INVES REAL NCF IRR	22.31	22.31	22.31	22.31	22.31	22.31	22.31	22.31	
GOVT CASH FLOWS									
GOVT NCF	23.18	24.80	26.53	28.39	30.38	32.51	34.78	37.22	326.24
REAL GOVT NCF	11.78	11.78	11.78	11.78	11.78	11.78	11.78	11.78	149.96
DIS5% GOVT REAL NCF	7.06	6.72	6.40	6.10	5.81	5.53	5.27	5.02	87.13
GOVT SHARE RNCF	78.55	78.55	78.55	78.55	78.55	78.55	78.55	78.55	80.18
TOTAL SURPLUS	29.51	31.57	33.78	36.15	38.68	41.39	44.28	47.38	412.99
REAL TOTAL SURPLUS	18.00	15.00	15.00	15.00	15.00	15.00	15.00	15.00	187.04
REAL PROJECT SURPLUS	7.30	7.30	7.30	7.30	7.30	7.30	7.30	7.30	89.90
PROJECT REAL IRR	33.73	33.73	33.73	33.73	33.73	33.73	33.73	33.73	

TABLE 3.3 *Malaysia: 51% government equity and 49% foreign investor equity, fully paid – gold price at US$450 an oz*

	1983	1984	1985	1986	1987	1988	1989	1990	1991	1992
OUTPUT				60000=0	60000=0	60000=0	60000=0	60000=0	60000=0	60000=0
PRICE	450.00	481.50	515.20	551.27	589.86	631.15	675.33	722.60	773.18	827.31
REVENUE				33.68	35.39	37.87	40.52	43.36	46.39	49.64
ACCUMULATED EXPL EXP	5.00									
DEVELOPMENT COSTS	5.00	16.05	11.45							
OPERATING COSTS				14.70	15.73	16.83	18.01	19.27	20.62	22.04
ROYALTIES				1.65	1.77	1.89	2.03	2.17	2.32	2.48
CAPITAL ALLOWANCE				37.50						
TAXABLE INCOME				−20.78	−2.88	16.26	20.48	21.92	23.45	25.09
INCOME TAX						6.50	8.19	8.77	9.38	10.04
DEVELOPMENT TAX						0.81	1.02	1.10	1.17	1.25
JOINT VENTURE DIVIDENDS				16.72	17.89	11.83	11.27	12.06	12.90	13.80
COMPANY DIVIDENDS				8.19	8.77	5.80	5.52	5.91	6.32	6.76
DIV WITHHOLDING TAX				1.23	1.32	0.87	0.83	0.89	0.95	1.01
GOVT DIVIDENDS				8.53	9.13	6.03	5.75	6.15	6.58	7.04
INVESTOR CASH FLOWS										
INVESTOR NCF	−4.90	−7.86	−5.61	6.96	7.45	4.93	4.69	5.02	5.37	5.75
REAL CASH FLOW	−4.90	−7.35	−4.90	5.69	5.69	3.51	3.13	3.13	3.13	3.13
DISD5% RCNCF	−4.78	−6.83	−4.34	4.79	4.56	2.69	2.28	2.17	2.67	1.97
INVESTOR REAL IRR	17.75	17.75	17.75	17.75	17.75	17.75	17.75	17.75	17.75	17.75

GOVT CASH FLOWS									
GOVT NCF	−5.10	−8.19	−5.84	11.41	12.21	16.11	17.82	19.07	20.40	21.83
REAL GOVT FLOWS	−5.10	−7.65	−5.10	9.30	9.31	11.49	11.87	11.87	11.87	11.87
DISD5% RGNCF	−4.98	−7.11	−4.51	7.85	7.48	8.78	8.65	8.23	7.84	7.47
GOVT SHARE RNCF				62.10	62.10	76.58	79.15	79.15	79.15	79.15
REAL GOVT CARRY INT FL	−5.10	−7.65	−5.10	6.96	6.96	4.30	3.83	3.83	3.83	3.83
GOVT CARRD INT IRR	21.38	21.38	21.38	21.38	21.38	21.38	21.38	21.38	21.38	21.38
TOTAL SURPLUS	−10.00	−16.05	−11.05	18.38	19.66	21.04	22.51	24.09	25.77	27.58
REAL TOTAL SURPLUS	−10.00	−15.00	−10.00	15.00	15.00	15.00	15.00	15.00	15.00	15.00
PROJECT CASH FLOW	−10.00	−16.05	−11.45	16.72	17.89	11.83	11.27	12.06	12.90	13.80
REAL PROJECT CASH FLOW	−10.00	−15.00	−10.00	13.65	13.65	8.43	7.51	7.51	7.51	7.51
PROJECT REAL IRK	21.38	21.38	21.38	21.38	21.38	21.38	21.38	21.38	21.38	21.38

	1993	1994	1995	1996	1997	1998	1999	2000	TOTAL
OUTPUT	60000b-0	60000b-0	60000b-0	60000b-0	60000b-0	60000b-	60000b-0	60000b-	89999*0
PRICE	885.22	947.18	1013.49	1084.43	1160.34	1241.56	1328.47	1421.47	
REVENUE	53.11	56.83	60.81	65.07	69.62	74.49	79.71	85.29	831.17
ACCUMULATED EXPL EXP									5.00
DEVELOPMENT COSTS									32.50
OPERATING COSTS	23.61	25.26	27.03	28.92	30.95	33.11	35.43	37.91	369.41
ROYALTIES	2.66	2.84	3.04	3.25	3.48	3.72	3.99	4.26	41.56
CAPITAL ALLOWANCE									37.50
TAXABLE INCOME	26.85	28.73	30.74	32.89	35.20	37.66	40.30	43.12	153.08
INCOME TAX	10.74	11.49	12.30	13.16	14.00	15.06	16.12	17.25	
DEVELOPMENT TAX	1.34	1.44	1.54	1.64	1.76	1.88	2.01	2.16	19.14
JOINT VENTURE DIVIDENDS	14.77	15.80	16.91	18.09	19.36	20.71	22.16	23.71	
COMPANY DIVIDENDS	7.24	7.74	8.29	8.87	9.49	10.15	10.86	11.62	121.51
DIV WITHOLDING TAX	1.09	1.16	1.24	1.33	1.42	1.52	1.63	1.74	18.23
GOVT DIVIDENDS	7.53	8.06	8.62	9.23	9.87	10.56	11.30	12.09	126.47
INVESTOR CASH FLOWS									
INVESTOR NCF	6.15	6.58	7.04	7.54	8.06	8.63	9.23	9.88	84.91
REAL CASH FLOW	3.13	3.13	3.13	3.13	3.13	3.13	3.13	3.13	
DISD5% RCNCF	1.87	1.78	1.70	1.62	1.54	1.47	1.40	1.33	35.26
INVESTOR REAL IRR	17.75	17.75	17.75	17.75	17.75	17.75	17.75	17.75	17.28

GOVT CASH FLOWS									
GOVT NCF	23.36	24.99	26.74	28.61	30.62	32.76	35.05	37.50	339.35
REAL GOVT FLOWS	11.87	11.87	11.87	11.87	11.87	11.87	11.87	11.87	154.74
DISD5% RGNCF	7.11	6.77	6.45	6.14	5.85	5.57	5.31	5.06	87.98
GOVT SHARE RNCF	79.15	79.15	79.15	79.15	79.15	79.15	79.15	79.15	81.44
REAL GOVT CARRY INT FL	3.83	3.83	3.83	3.83	3.83	3.83	3.83	3.83	46.32
GOVT CARRD INT IRR	21.38	21.38	21.38	21.38	21.38	21.38	21.38	21.38	424.26
TOTAL SURPLUS	29.51	31.57	33.78	36.15	38.68	41.39	44.28	47.38	
REAL TOTAL SURPLUS	15.00	15.00	15.00	15.00	15.00	15.00	15.00	15.00	190.06
PROJECT CASH FLOW	14.77	15.80	16.91	18.09	19.36	20.71	22.16	23.71	210.49
REAL PROJECT CASH FLOW	7.51	7.51	7.51	7.51	7.51	7.51	7.51	7.51	90.82
PROJECT REAL IRR	21.38	21.38	21.38	21.38	21.38	21.38	21.38	21.38	

TABLE 3.4 Sierra Leone: 31% government equity fully paid, 20% 'free' equity – gold price at US$450 an oz

	1983	1984	1985	1986	1987	1988	1989	1990	1991	1992
OUTPUT				60000•0	60000•0	60000•0	60000•0	60000•0	60000•0	60000•0
PRICE	450.00	481.50	515.20	551.27	589.86	631.15	675.33	722.60	773.18	827.31
REVENUE				33.08	35.39	37.87	40.52	43.36	46.39	49.64
ACCUMULATED EXPL EXP	5.00									
DEVELOPMENT COSTS	5.00	16.05	11.45							
OPERATING COSTS				14.70	15.73	16.83	18.01	19.27	20.62	22.06
ROYALTIES				1.65	1.77	1.89	2.03	2.17	2.32	2.48
CAPITAL ALLOWANCE				2.50	0.18	0.21	0.24	0.28	0.34	0.42
VENTURE DEBT REPAYMENT										
LOAN	7.00	11.23	8.01							
INTEREST	0.77	2.01	2.89	2.41	1.92	1.44	0.96	0.48		
LOAN REPAYMENT				4.37	4.37	4.37	4.37	4.37	4.37	
INT WITHHOLDING TAX	0.35	0.90	1.30	1.08	0.87	0.65	0.43	0.22		
TAXABLE INCOME	−0.77	−2.78	−5.66	6.15	15.79	17.50	19.28	21.15	23.11	24.68
INCOME TAX				2.77	7.10	7.87	8.68	9.52	10.40	11.11
SURTAX				0.42	1.07	1.18	1.30	1.43	1.56	1.67
JT VENTURE DIVD				6.76	3.42	4.27	5.17	6.12	7.12	12.32
INVESTOR DIVIDENDS				3.31	1.68	2.09	2.53	3.00	3.49	6.04
GOVT DIVIDENDS				3.45	1.75	2.18	2.64	3.12	3.63	6.29

INVESTOR CASH FLOWS											
INVES NCF	−3.38	−5.80	−4.85	1.82	0.92	1.15	1.39	1.65	1.92	3.32	
INVES REAL NCF	−3.38	−5.42	−4.24	1.49	0.70	0.82	0.93	1.03	1.12	1.81	
DISD5% INVES REAL NCF	−3.30	−5.04	−3.75	1.25	0.56	0.63	0.68	0.71	0.74	1.14	
INVES REAL NCF IRR	6.40	6.40	6.40	6.40	6.40	6.40	6.40	6.40	6.40	6.40	
GOVT CASH FLOWS											
GOVT NCF	−0.05	−0.12	−0.17	10.86	13.31	14.72	16.21	17.80	19.48	24.26	
REAL GOVT NCF	−0.05	−0.11	−0.15	8.86	10.15	10.49	10.80	11.08	11.34	13.19	
DIS5% GOVT REAL NCF	−0.05	−0.10	−0.13	7.47	8.15	8.02	7.87	7.69	7.49	8.30	
GOVT %SHARE RNCF				85.64	93.52	92.74	92.09	91.52	91.04	87.96	
REAL GOVT CARRD INT FLS	−3.34	−5.23	−3.88	2.82	1.84	1.91	1.97	2.03	2.07	2.08	
GOVT CARRD INT IRR	12.89	12.89	12.89	12.89	12.89	12.89	12.89	12.89	12.89	12.89	
REAL TOTAL SURPLUS	−3.42	−5.53	−4.39	10.35	10.85	11.31	11.73	12.11	12.45	15.00	
REAL PROJECT SURPLUS	−3.77	−6.37	−5.52	5.52	2.61	3.05	3.44	3.81	4.14	6.70	
PROJECT REAL IRR	23.22	23.22	23.22	23.22	23.22	23.22	23.22	23.22	23.22	23.22	

	1993	1994	1995	1996	1997	1998	1999	2000	TOTAL
OUTPUT	60000e0	60000e0	60000e0	60000e0	60000e0	60000e0	60000e0	60000e0	89999e1
PRICE	885.22	947.18	1013.49	1084.43	1160.34	1241.56	1328.47	1421.47	
REVENUE	53.11	56.83	60.81	65.07	69.62	74.49	79.71	85.29	831.17
ACCUMULATED EXPL EXP									5.00
DEVELOPMENT COSTS									32.50
OPERATING COSTS	23.61	25.26	27.03	28.92	30.94	33.11	35.43	37.91	369.41
ROYALTIES	2.66	2.84	3.04	3.25	3.48	3.72	3.99	4.26	41.56
CAPITAL ALLOWANCE	0.52	0.67	0.89	1.25	1.87	3.12	6.25	18.75	37.50
VENTURE DEBT REPAYMENT									
LOAN									26.25
INTEREST									12.88
LOAN REPAYMENT									26.25
INT WITHHOLDING TAX									5.80
TAXABLE INCOME	26.33	28.06	29.85	31.64	33.32	34.54	34.05	24.37	166.42
INCOME TAX	11.85	12.63	13.43	14.24	14.99	15.54	15.32	10.97	24.96
SURTAX	1.78	1.89	2.01	2.14	2.25	2.33	2.30	1.64	195.35
JT VENTURE DIVD	13.23	14.21	15.30	16.52	17.95	19.79	22.68	30.51	195.35
INVESTOR DIVIDENDS	6.48	6.96	7.49	8.09	8.80	9.70	11.11	14.95	95.72
GOVT DIVIDENDS	6.74	7.25	7.80	8.42	9.16	10.09	11.57	15.56	99.63

INVESTOR CASH FLOWS							
INVES NCF	3.56	3.83	4.12	4.45	4.84	5.33	6.11	8.22	38.62
INVES REAL NCF	1.81	1.82	1.83	1.85	1.88	1.93	2.07	2.60	10.65
DIS 5% INVES REAL NCF	1.09	1.04	0.99	0.96	0.92	0.91	0.93	1.11	1.57
INVES REAL NCF IRR	6.40	6.40	6.40	6.40	6.40	6.40	6.40	6.40	
GOVT CASH FLOWS								
GOVT NCF	25.94	27.74	29.66	31.70	33.84	36.05	38.17	39.16	378.55
REAL GOVT NCF	13.19	13.18	13.17	13.15	13.12	13.07	12.93	12.40	179.82
DIS 5% GOVT REAL NCF	7.90	7.52	7.16	6.81	6.47	6.13	5.78	5.28	107.75
GOVT %SHARE RNCF	87.92	87.87	87.80	87.68	87.49	87.11	86.20	82.65	
REAL GOVT CARRD INT FLS	2.08	2.09	2.11	2.12	2.16	2.22	2.38	2.99	20.43
GOVT CARRD INT IRR	12.89	12.89	12.89	12.89	12.89	12.89	12.89	12.89	
REAL TOTAL SURPLUS	15.00	15.00	15.00	15.00	15.00	15.00	15.00	15.00	190.47
REAL PROJECT SURPLUS	6.72	6.75	6.79	6.85	6.96	7.17	7.68	9.66	72.20
PROJECT REAL IRR	23.22	23.22	23.22	23.22	23.22	23.22	23.22	23.22	

TABLE 3.5 Canada: 100% fully paid foreign investor participation – gold price at US$450 an oz

	1983	1984	1985	1986	1987	1988	1989	1990	1991	1992
OUTPUT				60000=0	60000=0	60000=0	60000=0	60000=0	60000=0	60000=0
PRICE	450.00	481.50	515.20	551.27	589.86	631.15	675.33	722.60	773.18	827.31
REVENUE				33.08	35.39	37.87	40.52	43.36	46.39	49.64
ACCUMULATED EXPL EXP	5.00									
DEVELOPMENT COSTS	5.00	16.05	11.45							
OPERATING COSTS				14.70	15.73	16.83	18.01	19.27	20.62	22.06
CAPITAL ALLOWANCE				14.75	6.82	4.78	3.34	2.34	1.64	1.15
PROCESSING ALLOWANCE				1.37	1.47	1.57	1.68	1.80	1.92	2.06
EARNED DEPLETION ALL				4.59	4.92	5.26	5.63	5.70	3.80	2.53
RESOURCE ALLOWANCE				4.59	4.92	5.26	5.63	6.02	6.44	6.89
QUE MNG TAX ASS INCOME				7.41	13.28	14.21	15.20	16.59	20.05	22.98
QUEBEC MINING TAX				1.28	2.61	2.84	3.09	3.43	4.30	5.17
QUE CORP TAX ASS INCOME				−5.56	−2.56	3.19	7.91	10.02	13.89	17.00
QUEBEC CORP TAX						0.38	0.95	1.20	1.67	2.04
FED CORP TAXABLE INCOME				−5.56	−2.56	3.19	7.91	10.02	13.89	17.00
FED CORP TAX						1.15	2.85	3.61	5.00	6.12
INVESTOR DIVIDENDS				17.09	17.05	16.67	15.63	15.84	14.81	14.25
INVESTOR CASH FLOWS										
INVESTOR NCF	−10.00	−16.05	−11.45	17.09	17.05	16.67	15.63	15.84	14.81	14.25
REAL CASH FLOW	−10.00	−15.00	−10.00	13.95	13.01	11.89	10.41	9.86	8.62	7.75
DISCOVNTED 5% RINCF	−9.76	−13.94	−8.85	11.76	10.45	9.09	7.58	6.84	5.69	4.87
INVESTOR REAL IRR	23.32	23.32	23.32	23.32	23.32	23.32	23.32	23.32	23.32	23.32

GOVT CASH FLOWS										
TOTAL GOVT NCF				1.28	2.61	4.37	6.89	8.25	10.97	13.33
REAL GOVT FLOWS				1.05	1.99	3.11	4.59	5.14	6.38	7.25
DISCOUNTED 5% RGNCF				0.88	1.60	2.38	3.34	3.56	4.22	4.56
GOVT SHARE RNCF				6.98	13.26	20.77	30.59	34.23	42.55	48.34
GOVT SHARE DRNCF				6.98	13.26	20.77	30.59	34.23	42.55	48.34
QUEBEC GOVT NCF				1.28	2.61	3.22	4.04	4.64	5.97	7.21
QUEBEC GOVT RNCF				1.05	1.99	2.30	2.69	2.89	3.47	3.92
QUEBEC GOVT 5% DIS RNCF				0.88	1.60	1.76	1.96	2.00	2.29	2.47
QUE GOVT SURPLUS SHARE%				6.98	13.26	15.31	17.94	19.25	23.15	26.15
TOTAL SURPLUS	−10.00	−16.05	−11.45	18.38	19.66	21.04	22.51	24.09	25.77	27.58
REAL TOTAL SURPLUS	−10.00	−15.00	−10.00	15.00	15.00	15.00	15.00	15.00	15.00	15.00
PROJECT CASH FLOW	−10.00	−16.05	−11.45	17.09	17.05	16.67	15.63	15.84	14.81	14.25
REAL PROJECT CASH FLOW	−10.00	−15.00	−10.00	13.95	13.01	11.89	10.41	9.86	8.62	7.75
PROJECT REAL IRR	23.32	23.32	23.32	23.32	23.32	23.32	23.32	23.32	23.32	23.32

	1993	1994	1995	1996	1997	1998	1999	2000	TOTAL
OUTPUT	60000▬0	60000▬0	60000▬0	60000▬0	60000▬0	60000▬0	60000▬0	60000▬0	89999▬1
PRICE	885.22	947.18	1013.49	1084.43	1160.34	1241.56	1328.47	1421.47	
REVENUE	53.11	56.83	60.81	65.07	69.62	74.49	79.71	85.29	831.17
ACCUMULATED EXPL EXP									5.00
DEVELOPMENT COSTS									32.50
OPERATING COSTS	23.61	25.26	27.03	28.92	30.94	33.11	35.43	37.91	369.41
CAPITAL ALLOWANCE	0.80	0.56	0.39	0.28	0.19	0.13	0.09	0.22	37.50
PROCESSING ALLOWANCE	2.20	2.36	2.52	2.70	2.89	3.09	3.31	3.54	34.48
EARNED DEPLETION ALL	1.69	1.13	0.75	0.50	0.33	0.22	0.15	0.30	37.50
RESOURCE ALLOWANCE	7.38	7.89	8.45	9.04	9.67	10.35	11.07	11.85	115.44
QUE MNG TAX ASS INCOME	25.61	28.09	30.51	32.95	35.46	38.07	40.83	43.55	
QUEBEC MINING TAX	5.96	6.70	7.43	8.16	8.91	9.70	10.52	11.34	91.44
QUE CORP TAX ASS INCOME	19.64	21.99	24.19	26.33	28.48	30.68	32.97	35.02	
QUEBEC CORP TAX	2.36	2.64	2.90	3.16	3.42	3.68	3.96	4.20	32.56
FED CORP TAXABLE INCOME	19.64	21.99	24.19	26.33	28.48	30.68	32.97	35.02	
FED CORP TAX	7.07	7.92	8.71	9.48	10.25	11.05	11.87	12.61	97.68
INVESTOR DIVIDENDS	14.12	14.32	14.74	15.35	16.09	16.96	17.93	19.23	240.09

INVESTOR CASH FLOWS								
INVESTOR NCF	14.12	14.32	14.74	15.35	16.09	16.96	17.93	19.23	202.59
REAL CASH FLOW	7.18	6.80	6.55	6.37	6.24	6.15	6.07	6.09	91.94
DISCOUNTED 5% RINCF	4.30	3.88	3.56	3.30	3.08	2.89	2.72	2.59	50.04
INVESTOR REAL IRR	23.32	23.32	23.32	23.32	23.32	23.32	23.32	23.32	
GOVT CASH FLOWS								
TOTAL GOVT NCF	15.39	17.26	19.04	20.80	22.58	24.42	26.35	28.15	221.68
REAL GOVT FLOWS	7.82	8.20	8.45	8.63	8.76	8.85	8.93	8.91	98.06
DISCOUNTED 5% RGNCF	4.69	4.68	4.59	4.47	4.32	4.16	3.99	3.79	55.22
GOVT SHARE RNCF	52.14	54.66	56.36	57.54	58.39	59.02	59.50	59.41	51.61
GOVT SHARE DRNCF	52.14	54.66	56.36	57.54	58.39	59.02	59.50	59.41	52.46
QUEBEC GOVT NCF	8.32	9.34	10.33	11.32	12.33	13.38	14.48	15.54	124.00
QUEBEC GOVT RNCF	4.23	4.44	4.59	4.70	4.78	4.85	4.90	4.92	55.71
QUEBEC GOVT 5% DIS RNCF	2.53	2.53	2.49	2.43	2.36	2.28	2.19	2.09	31.87
QUE GOVT SURPLUS SHARE%	28.18	29.58	30.58	31.31	31.88	32.33	32.70	32.80	29.32
TOTAL SURPLUS	29.51	31.57	33.78	36.15	38.68	41.39	44.28	47.38	424.26
REAL TOTAL SURPLUS	15.00	15.00	15.00	15.00	15.00	15.00	15.00	15.00	190.00
PROJECT CASH FLOW	14.12	14.32	14.74	15.35	16.09	16.96	17.93	19.23	202.59
REAL PROJECT CASH FLOW	7.18	6.80	6.55	6.37	6.24	6.15	6.07	6.09	91.94
PROJECT REAL IRR	23.32	23.32	23.32	23.32	23.32	23.32	23.32	23.32	

TABLE 3.6 *Papua New Guinea: 100% fully paid foreign investor participation with capital allowance spread over life of mine – gold price at US$ 450 an oz*

	1983	1984	1985	1986	1987	1988	1989	1990	1991	1992
Output	—	—	—	60 000	60 000	60 000	60 000	60 000	60 000	60 000
Price	450.00	481.50	515.20	551.27	589.86	631.15	675.33	722.60	773.18	827.31
Revenue	—	—	—	33.08	35.39	37.87	40.52	43.46	46.39	49.64
Accumulated exploration expenditure	5.00									
Development costs	5.00	16.05	11.45	—	—	—	—	—	—	—
Operating costs				14.70	15.73	16.83	18.01	19.27	20.62	22.06
Royalties				0.41	0.44	0.47	0.51	0.54	0.58	0.62
Capital allowance				9.37	2.01	2.01	2.01	2.10	2.01	1.81
Taxable income				8.59	17.21	18.56	20.00	21.54	23.18	25.15
Income tax				2.86	5.74	6.18	6.66	7.18	7.73	8.38
Net assessable revenue	−10.00	−16.05	−11.45	15.10	13.48	14.38	15.34	16.37	17.47	18.57
Accumulated NAR	−10.00	−28.05	−45.11	−39.03	−33.35	−25.64	−15.43	−2.15	14.88	18.57
Additional profits tax									5.46	6.81
Investor dividends				15.10	13.48	14.38	15.34	16.37	12.01	11.76
Dividend withholding tax				2.26	2.02	2.16	2.30	2.46	1.80	1.76
Investor cash flows										
Investor NCF	−10.00	−16.05	−11.45	12.83	11.46	12.22	13.04	13.91	10.21	10.00
Real cash flow	−10.00	−15.00	−10.00	10.48	8.74	8.71	8.69	8.66	5.94	5.44
Discounted RCNCF	−9.76	−13.94	−8.85	8.83	7.02	6.65	6.33	6.01	3.92	3.42
Investor real IRR	16.91	16.91	16.91	16.91	16.91	16.91	16.91	16.91	16.91	16.91

Government cash flows										
Government NCF	—	—	—	5.54	8.20	8.82	9.47	10.17	15.57	17.58
Real government flows	—	—	—	4.52	6.26	6.29	6.31	6.34	9.06	9.56
Discounted RGNCF	—	—	—	3.81	5.02	4.81	4.60	4.39	5.98	6.01
Government share RNCF	—	—	—	30.15	41.71	41.90	42.08	42.24	60.40	63.74
Government share DRNCF	—	—	—	30.15	41.71	41.90	42.08	42.24	60.40	63.74
Total surplus	−10.00	−16.05	−11.45	18.38	19.66	21.04	22.51	24.09	25.77	27.58
Real total surplus	−10.00	−15.00	−10.00	15.00	15.00	15.00	15.00	15.00	15.00	15.00
Real project cash flow	−10.00	−15.00	−10.00	12.33	10.29	10.25	10.22	10.19	6.99	6.40
Project real IRR	20.47	20.47	20.47	20.47	20.47	20.47	20.47	20.47	20.47	20.47

	1993	1994	1995	1996	1997	1998	1999	2000	Total
Output	60 000	60 000	60 000	60 000	60 000	60 000	60 000	60 000	89 999
Price	885.22	947.18	1013.49	1084.43	1160.34	1241.56	1328.47	1421.47	—
Revenue	53.11	56.83	60.81	65.07	69.62	74.49	79.71	85.29	831.17
Accumulated exploration expenditure	—	—	—	—	—	—	—	—	5.00
Development costs	—	—	—	—	—	—	—	32.50	32.50
Operating costs	23.61	25.26	27.03	28.92	30.94	33.11	35.43	37.91	369.41
Royalties	0.66	0.71	0.76	0.81	0.87	0.93	1.00	1.07	10.39
Capital allowance	1.63	1.46	1.32	1.19	1.07	0.96	0.86	7.78	37.50
Taxable income	27.22	29.40	31.70	34.15	36.74	39.49	42.42	38.53	—
Income tax	9.07	9.80	10.57	11.38	12.25	13.16	14.14	12.84	137.94
Net assessable revenue	19.77	21.06	22.46	23.95	25.56	27.29	29.15	33.47	—
Accumulated NAR	19.77	21.06	22.46	23.95	25.56	27.29	29.15	33.47	—
Additional profits tax	7.25	7.72	8.23	8.78	9.37	10.01	10.69	12.27	86.61
Investor dividends	12.52	13.34	14.22	15.17	16.19	17.28	18.46	21.20	226.82
Dividend withholding tax	1.88	2.00	2.13	2.28	2.43	2.59	2.77	3.18	34.02

Investor cash flows								
Investor NCF	10.64	11.34	12.09	12.89	13.76	14.69	15.69	18.02	155.30
Real cash flow	5.41	5.39	5.37	5.35	5.34	5.32	5.31	5.70	64.86
Discounted RCNCF	3.24	3.07	2.92	2.77	2.63	2.50	2.38	2.43	31.58
Investor real IRR	16.91	16.91	16.91	16.91	16.91	16.91	16.91	16.91	—
Government cash flows								
Government NCF	18.86	20.23	21.69	23.25	24.92	26.69	28.59	29.36	268.96
Real government flows	9.59	9.61	9.63	9.65	9.66	9.68	9.69	9.30	125.14
Discounted RGNCF	5.75	5.49	5.23	4.99	4.76	4.54	4.33	3.96	73.68
Government share RNCF	63.93	64.09	64.22	64.33	64.42	64.50	64.57	61.97	—
Government share DRNCF	63.93	64.09	64.22	64.33	64.42	64.50	64.57	61.97	—
Total surplus	29.51	31.57	33.78	36.15	38.68	41.39	44.28	47.38	424.26
Real total surplus	15.00	15.00	15.00	15.00	15.00	15.00	15.00	15.00	190.00
Real project cash flow	6.37	6.34	6.31	6.29	6.28	6.26	6.25	6.71	82.48
Project real IRR	20.47	20.47	20.47	20.47	20.47	20.47	20.47	20.47	—

TABLE 3.7 Papua New Guinea: 100% foreign investor participation but with immediate capital write-off – gold price at US$ 450 an oz

	1983	1984	1985	1986	1987	1988	1989	1990	1991	1992
Output	—	—	—	60 000	60 000	60 000	60 000	60 000	60 000	60 000
Price	450.00	481.50	515.20	551.27	589.86	631.15	675.33	722.60	773.18	827.31
Revenue	—	—	—	33.08	35.39	37.87	40.52	43.36	46.39	49.64
Accumulated exploration expenditure	5.00	—	—	—	—	—	—	—	—	—
Development costs	5.00	16.05	11.45	—	—	—	—	—	—	—
Operating costs	—	—	—	14.70	15.73	16.83	18.01	19.27	20.62	22.06
Royalties	—	—	—	0.41	0.44	0.47	0.51	0.54	0.58	0.62
Capital allowance	—	—	—	37.50	—	—	—	—	—	—
Taxable income	—	—	—	−19.54	−0.32	20.25	22.00	23.54	25.19	26.96
Income tax	—	—	—	—	—	6.75	7.33	7.85	8.40	8.98
Net assessable revenue	−10.00	−16.05	−11.45	17.96	19.22	13.82	14.67	15.70	16.80	17.97
Accumulated NAR	−10.00	−28.05	−45.11	−36.17	−24.18	−15.20	−3.57	11.41	16.80	17.97
Additional profits tax	—	—	—	—	—	—	—	4.18	6.16	6.59
Investor dividends	—	—	—	17.96	19.22	13.82	14.67	11.51	10.64	11.38
Dividend withholding tax	—	—	—	2.69	2.88	2.07	2.20	1.73	1.60	1.71
Investor cash flows										
Investor NCF	−10.00	−16.05	−11.45	15.27	16.34	11.74	12.47	9.79	9.04	9.67
Real cash flow	−10.00	−15.00	−10.00	12.46	12.46	8.37	8.31	6.09	5.26	5.26
Discounted RCNCF	−9.76	−13.94	−8.85	10.51	10.01	6.40	6.05	4.23	3.48	3.31
Investor real IRR	18.15	18.15	18.15	18.15	18.15	18.15	18.15	18.15	18.15	18.15

Government cash flows										
Government NCF	—	—	—	—	3.11	3.32	9.29	10.04	14.30	16.73	17.90
Real government flows	—	—	—	—	2.54	2.54	6.63	6.69	8.91	9.74	9.74
Discounted RGNCF	—	—	—	—	2.14	2.04	5.07	4.87	6.18	6.43	6.13
Government share RNCF	—	—	—	—	16.91	16.91	44.18	44.61	59.37	64.92	64.92
Government share DRNCF	—	—	—	—	16.91	16.91	44.18	44.61	59.37	64.92	64.92
Total surplus	−10.00	−16.05	−11.45	18.38	19.66	21.04	22.51	24.09	25.77	27.58
Real total surplus	−10.00	−15.00	−10.00	15.00	15.00	15.00	15.00	15.00	15.00	15.00
Project cash flow	−10.00	−16.05	−11.45	17.96	19.22	13.82	14.57	11.51	10.64	11.38
Real project cash flow	−10.00	−15.00	−10.00	14.66	14.66	9.85	9.78	7.17	6.19	6.19
Project real IRR	22.10	22.10	22.10	22.10	22.10	22.10	22.10	22.10	22.10	22.10

	1993	1994	1995	1996	1997	1998	1999	2000	Total
Output	60 000	60 000	60 000	60 000	60 000	60 000	60 000	60 000	89 999
Price	885.22	947.18	1013.49	1084.43	1160.34	1241.56	1328.47	1421.47	—
Revenue	53.11	56.83	60.81	65.07	69.62	74.49	79.71	85.29	831.17
Accumulated exploration expenditure									5.00
Development costs									32.50
Operating costs	23.61	25.26	27.03	28.92	30.94	33.11	35.43	37.91	369.41
Royalties	0.66	0.71	0.76	0.81	0.87	0.93	1.00	1.07	10.39
Capital allowance									37.50
Taxable income	28.84	30.86	33.02	35.33	37.81	40.45	43.29	46.32	—
Income tax	9.61	10.29	11.01	11.78	12.60	13.48	14.43	15.44	137.94
Net assessable revenue	19.23	20.58	22.02	23.56	25.21	26.97	28.86	30.88	—
Accumulated NAR	19.23	20.58	22.02	23.56	25.21	26.97	28.86	30.88	—
Additional profits tax	7.05	7.55	8.07	8.64	9.24	9.89	10.58	11.32	89.28
Investor dividends	12.18	13.03	13.94	14.92	15.96	17.08	18.28	19.56	224.15
Dividend withholding tax	1.83	1.95	2.09	2.24	2.39	2.56	2.74	2.93	33.62
Investor cash flows									
Investor NCF	10.35	11.08	11.85	12.68	13.57	14.52	15.53	16.62	153.03
Real cash flow	5.26	5.26	5.26	5.26	5.26	5.26	5.26	5.26	65.33
Discounted RCNCF	3.15	3.00	2.86	2.72	2.59	2.47	2.35	2.24	32.82
Investor real IRR	18.15	18.15	18.15	18.15	18.15	18.15	18.15	18.15	—

Government cash flows									
Government NCF	19.16	20.50	21.93	23.47	25.11	26.87	28.75	30.76	271.24
Real government flows	9.74	9.74	9.74	9.74	9.74	9.74	9.74	9.74	124.67
Discounted RGNCF	5.83	5.56	5.29	5.04	4.80	4.57	4.35	4.15	72.44
Government share RNCF	64.92	64.92	64.92	64.92	64.92	64.92	64.92	64.92	—
Government share DRNCF	64.92	64.92	64.92	64.92	64.92	64.92	64.92	64.92	—
Total surplus	29.51	31.57	33.78	36.15	38.68	41.39	44.28	47.38	424.26
Real total surplus	15.00	15.00	15.00	15.00	15.00	15.00	15.00	15.00	190.00
Project cash flow	12.18	13.03	13.94	14.92	15.96	17.08	18.28	19.56	186.65
Real project cash flow	6.19	6.19	6.19	6.19	6.19	6.19	6.19	6.19	83.03
Project real IRR	22.10	22.10	22.10	22.10	22.10	22.10	22.10	22.10	—

Appendix 4

DEPRECIATION METHODS

The effects of the three principal methods of depreciation – straight line, sum-of-years' digits and double declining balance – are considered here. To illustrate these effects it is assumed that an asset is purchased for US$2 500 and has an estimate useful life of six years with a US$400 salvage value after six years. (The salvage value is the amount that can be realised from selling the asset after its useful life has ended.)

With the straight line method, a uniform annual depreciation charge of US$350/year is allowed. This figure is arrived at by simply dividing the economic life into the total cost of the machine minus the estimate salvage value:

$$\frac{\text{US\$2 500 cost} - \text{US\$400 salvage value}}{6 \text{ years}} = \text{US\$350/year depreciation charge}$$

If the estimated salvage value had been less than 10 per cent of the original cost, it could have been ignored.

The double declining balance (DDB) method of depreciation requires the application of a constant rate of depreciation each year to the undepreciated value of the asset at the close of the previous year. To calculate DDB depreciation we first find the fraction $1/N$, where N is the life of the asset. In our example, $1/N = 1/6 = 0.1667$. This fraction is then doubled, giving 0.3333 in this case, and it is called the 'depreciation rate'. This rate is applied to the full purchase price of the machine, not to the cost less salvage value. Therefore, depreciation under the DDB method during the first year is calculated as follows:

$$0.3333 \, (\text{US\$2 500}) = \text{US\$833}$$

Depreciation during the second year is calculated by applying the 33.33 per cent rate (or 0.3333) to the undepreciated balance as follows:

$$0.3333 \, (\text{US\$2 500} - \text{US\$833}) = 0.3333 \, (\text{US\$1 667}) = \text{US\$556}$$

The process is continued for other years, until the total depreciation taken equals the cost of the asset less the estimated salvage value. Thus, in the case illustrated here, the asset is fully depreciated during the fifth year.

Under the sum-of-years' digits (SYD) method, the yearly depreciation allowance is determined as follows:

Calculate the SYDs. In our example there is a total of 21 digits: $1+2+3+4+5+6 = 21$. This figure can also be arrived at by means of the sum of an algebraic progression equation, where N is the life of the asset:

$$\text{Sum} = N\left(\frac{N+1}{2}\right)$$

$$= 6\left(\frac{6+1}{2}\right) = 21$$

Divide the number of remaining years by the SYDs and multiply this fraction by the depreciable cost (total cost minus salvage value) of the asset:

Year 1: $\left(\frac{6}{21}\right)$ (US$2 100) = US$600 depreciation

Year 2: $\left(\frac{5}{21}\right)$ (US$2 100) = US$500 depreciation

Year 6: $\left(\frac{1}{21}\right)$ (US$2 100) = US$100 depreciation

The total amount of depreciation taken under the DDB method is equal to that under the straight line method, but under DDB the depreciation is taken faster, as it is under sum-of-years' digits (see Table A4.1). Thus, DDB and sum-of-years' digits are in a sense accelerated depreciation methods.

TABLE A4.1 *The three depreciation methods and a comparison of the depreciation charges under each method over a six year period.*

Year	Straight Line (US$)	Double Declining Balance (DDB) (US$)	Sum-of-Years' Digits (SYD) (US$)
1	350	833	600
2	350	556	500
3	350	370	400
4	350	247	300
5	350	94[a]	200
6	350	—	100
Total	2 100	2 100	2 100

[a] The maximum depreciation that can be taken is the value of cost minus salvage, or US$2100 in this example. Thus, US$94 of depreciation in year 5 exhausts the depreciation allowed under DDB.
Source: Kumar and Walrond (1983), p. 166.

Figure A4.1 illustrates diagrammatically the straight-line, double declining balance and sum-of-years' digits methods.

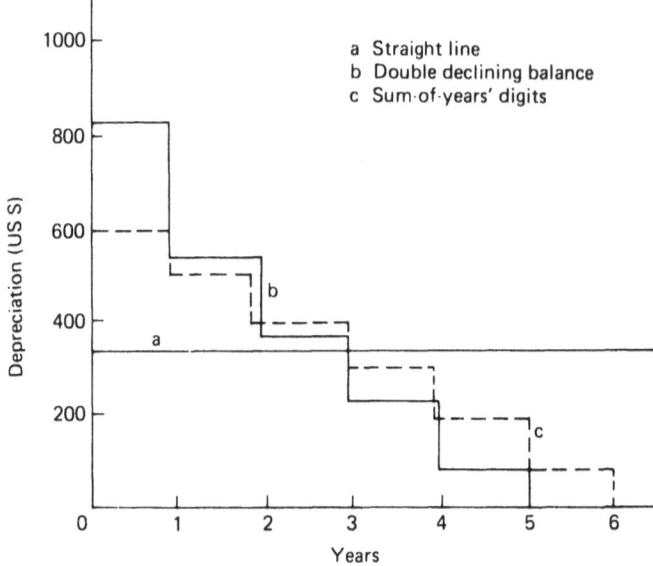

FIGURE A4.1 *Depreciation methods compared*

Appendix 5

NUMERICAL EXAMPLE OF A CASH FLOW BASED RENT RESOURCE TAX

The mode of operation of a rent resource tax (RRT) varies from regime to regime. The example presented below is a 'hybrid' form where income tax and royalty are elements of the fiscal package, and where the RRT is an item that could be deducted for income tax in the year in which it is paid. In other words the RRT is in some sense an additional royalty that is profit-based. The example that is chosen is not theoretically 'pure' but rather reflects what is becoming *practically* acceptable, taking into account the credibility of the tax in the United States. The formula to calculate the RRT may be written as follows:

$$\left(\frac{ER_{t-1}}{ER_t}\right) A_t (100\% + R_t) + B_t$$

(where ER_t is the mean of the average buying and selling rate of the local currency against the foreign currency (usually US dollars) during the year 't', that is, the year for which the calculation is made and

ER_{t-1} is the ER_t for the preceding year of income for which the calculation is made.

A_t is accumulated net cash receipts at the end of year $t-1$.

R_t is the 'threshold' or 'accumulation' rate which is defined *a priori*.

B_t is the net cash receipts for the year of income for which the calculation is to be made.)

Practical points to note

(a) The net cash receipts are usually clearly stipulated, including *actual* cash flows. Interest is not usually included for both theoretical and practical reasons. Theoretically it is presumed that the accumulation rate incorporates the interest element, while practically the aim is to minimise transfer pricing. Note that depreciation is an accounting concept and does not directly form part of the net cash receipts.
(b) In defining exchange rates, it is necessary to spell out precisely which ones are meant because of the numerous rates that prevail during the day. It is

TABLE A5.1 Numerical example of calculating a rent resource tax that is deductible for income tax purposes

	1	2	3	4	5	6	7	8	9
Revenue	0.0	0.0	100.0	150.0	160.0	165.0	150.0	170.0	180.0
Capital cost	50.0	100.0	0.0	.0	0.0	0.0	50.0[a]	0.0	0.0
Operating costs[b]	0.0	0.0	10.0	15.0	20.0	30.0	42.0	50.0	55.0
Royalty[c]	0.0	0.0	10.0	15.0	16.0	16.5	15.0	17.0	18.0
Capital allowance[d]	0.0	0.0	30.0	30.0	30.0	30.0	40.0	10.0	10.0
Taxable income[e]	0.0	0.0	50.0	90.0	94.0	77.0	14.9	93.0	64.7
Income tax[f]	0.0	0.0	20.0	36.0	37.6	30.8	6.0	37.2	25.9
Net cash receipts[g] for RRT	−50.0	−100.0	60.0	84.0	86.4	76.2	−1.1	65.8	48.8
Accumulated net[h] cash receipts for RRT	−50.0	−157.50	−121.1	−55.2	22.9	76.2[i]	−1.1[j]	64.5	48.8
RRT assessed[k]	0.0	0.0	0.0	0.0	11.5	38.1	0.0	2.3	24.4
RRT paid[l]	0.0	0.0	0.0	0.0	0.0	11.5	38.1	0.0	32.3
Govt cash[m] flow	0.0	0.0	30.0	51.0	53.6	58.8	59.0	54.2	76.2
Investor cash[n] flow	−50.0	−100.0	60.0	84.0	86.4	76.2	−1.1	65.8	48.8

[a] This represents new capital investment, say an expansion.
[b] Includes working capital and any allowable interest payments.
[c] 10 per cent of gross revenue.
[d] 5-year straight-line (note: this is not a cash flow but an accounting calculation for income tax).

e Taxable income revenue = operating costs − royalty − capital allowance − RRT paid
f 40 per cent of taxable income.
g Net cash flow = revenue − capital costs − operating cost − royalty − income tax − RRT paid.
h The net cash receipts are accumulated at 15 per cent 'threshold rate' and adding to this the current net cash flows. Note that accumulation of the net cash receipts by the threshold rate return only proceeds as long as the net cash receipts are negative, allowing the investor to recoup his original investment as well as earn the 15 per cent rate of return before paying the rent resource tax.
i The positive accumulated value shows that the 15 per cent rate of return has been realised and the excess is taxed at 50 per cent.
j Accumulation takes place in the following year as the value is negative.
k 50 per cent of positive accumulated value.
l This is the RRT paid based on the previous year's assessment. Note that in this example RRT is deductible for income tax purposes in the year it is actually paid, that is, the year following the year of assessment.
m Government cash flow = royalty + income tax + RRT paid.
n Investor cash flow = revenue − capital costs − operating costs − royalty − income tax − RRT paid.

useful to list a particular publication, that is, IMF Financial Statistics, and specify the exchange rate that is published for that particular month or year.

(c) The choice of accumulation could vary, but has been based on a fixed predetermined amount (say 15 per cent) or on the US prime rate or US AAA bond rate, or London Interbank Offer Rate (LIBOR) or any combination agreed upon. The rate usually remains fixed throughout the agreement. There could be a progressive accumulation rate scheme attracting different rent resource tax rates, that is, 50 per cent after a 15 per cent rate of return is achieved, 60 per cent after 20 per cent, 70 per cent after 25 per cent and so on.

In the example, the following are assumed:

(i) Income tax rate = 40 per cent;
(ii) RRT rate = 50 per cent of positive accumulated net cash receipts;
(iii) Capital allowance = 5-year straight line;
(iv) Royalty = 10 per cent of gross revenue;
(v) Threshold rate of return = 15 per cent.

Appendix 6

EXPLANATION OF SOME BASIC FINANCIAL CONCEPTS

(1) Compound Value or Future Value

Suppose one has $100 in a savings bank that pays 7 per cent interest compounded annually. How much will one have at the end of five years? The terms may be defined as follows:
 PV = present value of the account, that is, the initial amount of $100.
 i = interest rate paid by the bank (in this case 7 per cent).
 I = dollars of interest earned at the end of the year.
 FV_n = future value, or ending amount, of the account at the end of n years. (PV is the value at the present time, where FV_n is the value n years into the future (in this case five years), after compound interest has been earned.)

The calculations are as follows:

Year	Initial amount (PV)	$\times (1+i)$	= End amount (FV_n)
1	$100.00	1.07	$107.00
2	$107.00	1.07	$114.49
3	$114.49	1.07	$122.50
4	$122.50	1.07	$131.08
5	$131.08	1.07	$140.26

With a good calculator it is easy to work out $(1+i)^n$ directly. However, there are tables for values of $(1+i)^n$ for wide ranges of i and n (see extract in Table A6.1).

In equation form, the future value of $1 at the end of n periods is:
$FV_n = PV(1+i)^n$

The higher the rate of interest, the faster the rate of growth (see diagram following):

Note that while compounding is often on an annual basis, it can be monthly, quarterly, semi-annually or for any other period. For example, to do semi-annual compounding, which means that interest is actually paid each six months, the annual interest rate is divided by two, but twice as many

TABLE A6.1 *Future and present value tables*
(Future Value of $1 at the end of n Periods $= (1+1)^n$)

Period	1%	2%	3%	4%	5%	6%	7%	8%	9%	10%
1	1.0100	1.0200	1.0300	1.0400	1.0500	1.0600	1.0700	1.0800	1.0900	1.1000
2	1.0201	1.0404	1.0609	1.0816	1.1025	1.1236	1.1449	1.1664	1.1881	1.2100
3	1.0303	1.0612	1.0927	1.1249	1.1576	1.1910	1.2250	1.2597	1.2950	1.3310
4	1.0406	1.0824	1.1255	1.1699	1.2155	1.2625	1.3108	1.3605	1.4116	1.4641
5	1.0510	1.1041	1.1593	1.2167	1.2763	1.3382	1.4026	1.4693	1.5386	1.6105
6	1.0615	1.1262	1.1941	1.2653	1.3401	1.4185	1.5007	1.5869	1.6771	1.7716
7	1.0721	1.1487	1.2299	1.3159	1.4071	1.5036	1.6058	1.7138	1.8280	1.9487
8	1.0829	1.1717	1.2668	1.3686	1.4775	1.5938	1.7182	1.8509	1.9926	2.1436
9	1.0937	1.1951	1.3048	1.4233	1.5513	1.6895	1.8385	1.9990	2.1719	2.3579
10	1.1046	1.2190	1.3439	1.4802	1.6289	1.7908	1.9672	2.1589	2.3674	2.5937
11	1.1157	1.2434	1.3842	1.5395	1.7103	1.8983	2.1049	2.3316	2.5804	2.8531
12	1.1268	1.2682	1.4258	1.6010	1.7959	2.0122	2.2522	2.5182	2.8127	3.1384
13	1.1381	1.2936	1.4685	1.6651	1.8856	2.1329	2.4098	2.7196	3.0658	3.4523
14	1.1495	1.3195	1.5126	1.7317	1.9799	2.2609	2.5785	2.9372	3.3417	3.7975
15	1.1610	1.3459	1.5580	1.8009	2.0789	2.3966	2.7590	3.1722	3.6425	4.1772

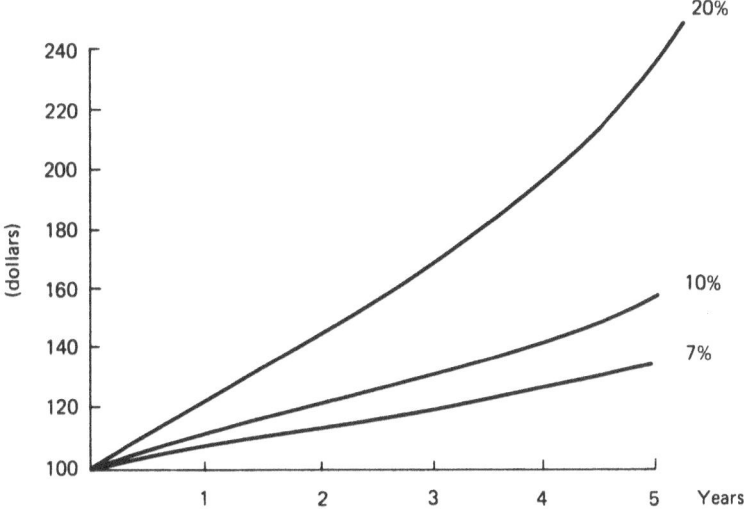

compounding periods are used, because interest is paid twice a year. At 7 per cent interest semi-annually the $100 at the end of one year will earn $107.12 compared to $107.00 on an annual basis.

(2) Present Value and the Discounting Process

Generally speaking, the present value of a sum due n years in the future is the amount which, if it were on hand today, would grow to equal the future sum. Since $100 would grow to $140.26 in five years at a 7 per cent interest rate, $100 is defined as the present value of $140.26 due five years in the future when the appropriate discount rate is 7 per cent. In other words, finding present values or discounting is simply the reverse of compounding. The future value formula presented above can readily be transformed into a present value formula as follows:

$FV_n = PV(1 + i)^n$ which, when solved for PV, gives

$$PV = \frac{FV_n}{(1+i)^n} = FV_n \left[\frac{1}{(1+i)}\right]^n$$

Tables have been constructed to determine the present values of $1 due at the end of n periods (an extract is presented in Table A6.2).

What is the present value of $140.26 at the end of five years at a discount rate of 7 per cent? Turning to Table A6.2, look down the 7 per cent columns to the fifth

TABLE A6.2 Future and present value tables
(Present Value of $1 = 1/(1+1)^n$)

Period	1%	2%	3%	4%	5%	6%	7%	8%	9%	10%
1	.9901	.9804	.9709	.9615	.9524	.9434	.9346	.9259	.9174	.9091
2	.9803	.9612	.9426	.9246	.9070	.8900	.8734	.8573	.8417	.8264
3	.9706	.9423	.9151	.8890	.8638	.8396	.8103	.7938	.7722	.7513
4	.9610	.9238	.8885	.8548	.8227	.7921	.7629	.7350	.7084	.6830
5	.9515	.9057	.8626	.8219	.7835	.7473	.7130	.6806	.6499	.6209
6	.9420	.8880	.8375	.7903	.7462	.7050	.6663	.6302	.5963	.5645
7	.9327	.8706	.8131	.7599	.7107	.6651	.6227	.5835	.5470	.5132
8	.9235	.8535	.7894	.7307	.6768	.6274	.5820	.5403	.5019	.4665
9	.9143	.8368	.7664	.7026	.6446	.5919	.5439	.5002	.4604	.4241
10	.9053	.8203	.7441	.6756	.6139	.5584	.5083	.4632	.4224	.3855
11	.8953	.8043	.7224	.6496	.5847	.5268	.4751	.4289	.3875	.3505
12	.8874	.7885	.7014	.6246	.5568	.4970	.4440	.3971	.3555	.3186
13	.8787	.7730	.6810	.6006	.5303	.4688	.4150	.3677	.3262	.2897
14	.8700	.7579	.6611	.5775	.5051	.4423	.3878	.3405	.2992	.2633
15	.8613	.7430	.6419	.5553	.4810	.4173	.3624	.3152	.2745	.2394

row (that is, Period = 5). The figure shown is 0.7130, which can be labelled as the present value interest factor (PVIF) to determine the present value of $140.26 payable in five years, discounted at 7 per cent:

$$PV = FV_5 \, (PVIF_{k,n})$$
$$= \$140.26 \, (0.7130)$$
$$= \$100$$

where k is the appropriate discount rate.

The diagram following shows the relationship between present value, interest rate and time:

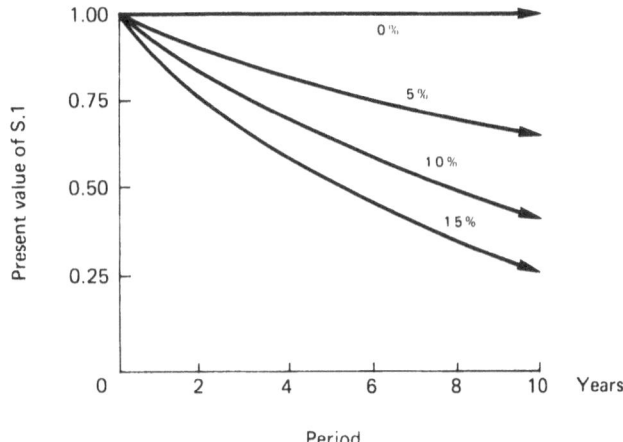

Even at relatively low discount rates, the present value of funds due in the distant future are quite small. For example $1.00 due in ten years is worth about 61 cents today, if the discount rate is 5 per cent, but is only worth 25 cents today at a 15 per cent discount rate. Similarly, $1.00 due in five years at 10 per cent is worth 62 cents today, but at the same discount rate $1.00 due in ten years is worth only 39 cents today.

(3) **Discounted Cash Flow Technique**

The *discounted cash flow technique* (DCF) uses the net present value method by finding the present value of the expected net cash flows of an investment, discounted at the appropriate percentage rate, and subtracted from it the initial cost outlay of the project. In equation form:

$$\text{NPV} = \sum_{t=1}^{n} \left(\frac{R}{(1+i)}t - C \right)$$

(where R represents the net annual receipts or net cash flows, i is the discount rate, or the project's cost of capital, and n is the project's expected life. C is the initial cost of the project).

The NPV is used as an indicator of the viability of the project. A positive NPV signifies that the investment expenditures will earn a higher return than that specified by the discount rate, and would therefore show an acceptable project. If the NPV is equal to zero, the investment will earn the exact return specified by the discount rate, and in some sense may signify a marginal project. If the NPV is negative, the investment does not earn as large a return as specified by the discount rate, and might suggest an unacceptable project from the financial point of view.

(4) The Internal Rate of Return (IRR)

The internal rate of return (IRR) is the discount rate that equates the present value of the expected future cash flows to the initial cost of the project. The equation for calculating this rate is:

$$\sum_{t=1}^{n} \frac{R_i}{(1+i)^t} - C = 0$$

In this equation the one unknown value is i, the discount rate. Some value of i will cause the sum of the discounted net cash flows to equal the initial cost of capital, making the equations equal to zero. The value of i is defined as the internal rate of return. Notice that the IRR formula is simply the NPV formula solved for the particular discount rate that causes the NPV to equal zero. In other words, the same basic equation is used for both methods, but in the NPV method the discount rate, i, is specified and the NPV is found, while in the IRR method the NPV is specified equal to zero and the value of i that forces NPV to equal zero is found. Most computer financial packages and some sophisticated calculators can compute the IRR in seconds. Manual calculations will have to be done by trial and error by going through various discount rates.

If the IRR exceeds the cost of the funds used to finance the project, a surplus is left over after paying for the cost of capital. If the IRR is below the cost of capital, the project imposes a cost on shareholders and the project may not be undertaken. The interest in the IRR lies in this 'breakeven' characteristic.

There are various reservations about the IRR method expressed in the literature, as well as the comparability of the NPV with the IRR method. For these the reader is referred to Brigham (1979), Levy and Sarnat (1982) and Bromwich (1976).

Bibliography

Anders, G. (1980) *Economics of Mineral Extraction* (New York: Praeger).
Asante, S. K. B. (1979) 'Stability of Contractral Relations in the Transnational Investment Process', *International and Comparative Law Quarterly*, vol. 28, pp. 401–00.
Baumol, W. J. and W. E. Oates (1976) 'Conservation of resources and the price system', in Leister, R. D. and S. L. Friedlander (eds) *Economics of Resources*, vol. 3, pp. 15–34.
Bennett, H. J. (1973) 'The mineral policy of selected foreign nations considering direct foreign investment in the mineral industry', *Quarterly of the Colorado School of Mines*, vol. 68, no. 4, pp. 213–59.
Bonis, G. (1972) 'Recommendations for the refocussing of international assistance programs in natural resources', *Earth Science Aid to Developing Countries*, Symposium 2, 24th Int. Geol. Cong., Montreal, pp. 61–6.
Bosson, R. and B. Varon (1977) *The Mining Industry and the Developing Countries*. (A World Bank Research Publication, Oxford University Press) p. 292.
Brewster, H. (1973) 'Economic dependence: A quantitative interpretation' (A Special Issue of Social and Economic Studies), *Institute of Social and Economic Research*, vol. 22, no. 1, 90–5.
Brigham, E. F. (1979) *Financial Management: Theory and Practice* (Hinsdale, Ill.: Dryden Press).
Bromwich, M. (1976) *The Economics of Capital Budgeting* (Harmondsworth: Penguin).
Brown, R. D. (1978) 'Canadian mineral taxation – a brief perspective', *CIM Bulletin*.
Campbell, H. F., Rainer, W. D. and A. Scott (1976) 'Resource rent: How much and for whom?', in Scott, A. (ed.) *Natural Resource Revenue: A Test of Federalism* (British Columbia Institute for Economic Policy Analysis series) pp. 118–36.
Carman, J. S. (1972) 'Limiting factors in foreign development aid with particular reference to mineral exploration', *Earth Science Aid to Developing Countries*, Symposium 2, 24th Int. Geol. Cong., Montreal, pp. 189–96.
Chandler, A. T. (1980) 'Role of mining legislation in mineral development strategies: a lawyer's point of view', *East-West Centre Workshop on Mineral Policies to Achieve Development objectives*, Honolulu.

Church, A. M. (1981) *Taxation of Non-renewable Resources* (Lexington, Mass.: Lexington Books).
Ciriachy-Wantrup, S. V. (1959) 'Economics and policies of resource conservation', in Huberty and Flock (eds) *Natural Resources*, pp. 500–26.
Cobbe, J. H. (1979) *Governments and Mining Companies in Developing Countries* (Boulder, Colorado: Westview).
Commonwealth Secretariat (1977a) 'Mining exploration and exploitation – proposals for a fiscal regime'.
Commonwealth Secretariat (1977b) 'Some policy and legal issues affecting mining legislation and agreements in african commonwealth countries', Technical Assistance Group.
Commonwealth Secretariat (1977c) 'Taxation in the mining industry', Technical Assistance Group.
Commonwealth Secretariat (1977d) 'The legislative framework, agreements and financial impositions affecting the mining industry in african commonwealth countries', Technical Assistance Group.
Commonwealth Secretariat (1978) 'Workshops on mining legislations and mineral resources agreement', Gaborone, Botswana.
Commonwealth Secretariat (1981) 'Application of new legal techniques to mining agreements in developing countries'.
Commonwealth Secretariat (1981) 'The legislative framework, agreements and financial impositions affecting the mining industry in pacific commonwealth countries', Technical Assistance Group.
Conrad, R. F. and B. R. Hool (1981) *Taxation of Mineral Resources* (Lexington, Mass.: Lexington Books).
Dales, J. H. (1960) 'Comment on connections between natural resources and economic growth', in Spengler, J. J. (ed.) *Natural Resources and Economic Growth* (University of Michigan) pp. 16–19.
Dundas, Carl (1981) 'The Development and Utilisation of Minerals in Africa: Management and Policies', Commonwealth Secretariat, London.
Ely, N. (1961, 1970–74) 'Summary of mining and petroleum laws of the world', US Bureau of Mines Information Circulars: IC 8017, Washington, D.C. 1961; IC 8482 Western Hemisphere 1970; IC 8514 East Asia and the Pacific 1971; IC 8610 Africa 1974; Vols I, II, III, IV, V, Washington, D.C. 1970–74.
Emerson, C. (1980) 'Taxing natural resource projects', *Natural Resources Forum*, vol. 4, no. 2, pp. 123–45.
Erickson, G. K. (1977) 'Work commitment bidding', in M. Crommelin and A. R. Thompson (eds) *Mineral Leasing as an Instrument of Public Policy* (Vancouver: University of British Columbia).
Faber, M. (1982) 'Some old and new devices in mineral royalties and taxation', in *Legal and Institutional Arrangments in Minerals Development*, Mining Journal Books, pp. 104–22.
Faber, M. and R. Brown (1980) 'Changing the rules of the game: political risk, instability and fair play in mineral concession contracts', *Third World Quarterly*, vol. 2, no. 1, pp. 100–17.

Francis, A. A. (1982) 'Transfer prices, tax minimisation and the US transnational corporation', *IBA Review*, vol. 8.
Freeman, P. (1980) 'Ambiguities in the operational meanings of free equity and additional profits tax', *East-West Centre Workshop on Mineral Policies to Achieve Development Objectives*, Honolulu.
Fritzche, M. and M. Bartels *et al.* (1981) *Bibliography on Transnational Law of Natural Resources*, Studies in Transnational Law and Natural Resources (Frankfurt am Main: Alfred Metzner Verlag).
Gaffney, M. (1967) 'Editor's conclusions', in Gaffney, M. (ed.) *Extractive Resources and Taxation* (University of Wisconsin Press) pp. 333–419.
Garnaut, R. and A. R. Ross (1975) 'Uncertainty, risk aversion and the taxing of natural resource projects', *Economic Journal*, vol. 85, no. 338, pp. 272–87.
Garnaut, R. and A. C. Ross (1977) 'A new tax for natural resources projects. Mineral leasing as an instrument for public policy', in M. Crommelin and A. R. Thompson (eds) *Mining Leasing as an Instrument of public policy* (Vancouver: University of British Columbia).
Garnaut, R. and A. C. Ross (1980) 'Relationships between governments and investors', *East-West Centre Workshop on Mineral Policies to Achieve Development Objectives*, Honolulu (June 9–13) (mimeograph).
Gillis, M. and L. T. Wells (1980) *Tax and Investment Policies for Hard Minerals* Cambridge, Mass.: Ballinger).
Girvan, N. (1970) 'Multinationals, Corporations and Dependent-Under-Development in Mineral Export Economies', *Social and Economic Studies*, vol. 19, no. 4, pp. 490–526.
Girvan, N. (1976) *Corporate Imperialism: Conflict and Expropriation. Transational Corporations and Economic Nationalism in the Third World* (New York: M. E. Sharpe Inc.).
Girvan, N. (1971) 'The denationalization of Caribbean bauxite: Alcan in Guyana', *New World Quarterly*, vol. 5, no. 3, pp. 35–48.
Girvan, N. (1972) 'The Guyana-Alcan conflict and the Nationalization of Demba', *New World Quarterly*, vol. 5, no. 4, pp. 38–49.
Gordon, P. A. L. (1971) 'Mineral royalties', *Trans. Inst. Min. Metall.*, vol. 80, A32-A37.
Gordon, P. A. L. (1971) 'Mining royalties–authors reply to discussion', *Trans. Inst. Min. Metall.*, vol. 80, A140-A141.
Govett, M. H. and H. A. Robinson (comp.) (1980) *Tin* (Sydney: Australian Mineral Economics Private Ltd).
Herfindahl, O. C. (1967) 'Depletion and economic theory', in Gaffney, M. (ed.) *Extractive Resources and Taxation* (University of Wisconsin Press) pp. 63–90.
Herfindahl, O. C. and A. V. Kneese (1974) *Economic Theory of Natural Resources* (Resources for the Future Inc.) p. 405.
Hogan, J. D. (1967) 'Resource exploitation and optimum tax policies: A control model approach, in Gaffney, M. (ed.) *Extractive Resources and Taxation* (University of Wisconsin Press) pp. 91–108.

Johnson, C. J. (1980) 'Taking the take but not the risk', *East-West Centre Workshop on Mineral Policies to Achieve Development Objectives*, Honolulu.
Johnson, C. J. (1981) 'Mineral objectives, policies and strategies in Botswana – analysis and lessons', *Natural Resources Forum*, vol. 5, pp. 347–67.
Kirchner, C. et al. (1979) *Mining Ventures in developing countries, Interests, Bargaining Process and Legal Concepts, Studies in Transnational Law and Natural Resources* (vol. 1) (Frankfurt am Main: Alfred Metzner Verlag).
Kumar, R. (1982) 'An introduction to computer assisted financial modelling of mining projects for developing countries', Technical Assistance Group, Commonwealth Secretariat, London.
Kumar, R. and G. Walrond (1983) 'Economic implications of capital allowance schemes for mining projects in developing countries', *Resource Policy*, pp. 155–68.
Laming, D. J. C. and D. E. Ajakalye (eds) (1977) 'New directions in mineral development policies', *Ass. Geosc. Inter. Dev.*
Lang, A. G. and M. Crommelin (1979) *Australian Mining and Petroleum Laws – An Introduction* (Melbourne: Butterworths).
Lanning, G. with M. Muller (1979) *Africa Undermined* (Harmondsworth: Penguin Books).
Levy, H. and M. Sarnat (1982) *Capital Investment and Financial Decisions* (Englewood Cliffs, New Jersey: Prentice-Hall).
Lipton, C. J. (1976) 'Government negotiating techniques and strategies', paper presented at the *Inter-regional Workshop on Negotiation and Drafting of Mining Development Agreements* (Buenos Aires, November 1973) reprinted in *Mining Journal Books*, London.
Mackenzie, B. W. (1983) *Mineral Project Evaluation Techniques and Applications*, Seminar Notes, Department of Mining and Metallurgical Engineering, McGill University, Montreal, Quebec.
Mackenzie, B. W. and M. L. Bilodeau (1979) *The Effect of Taxation on Base Metal Mining in Canada*, Centre for Research Studies (Kingston, Ontario: Queen's University).
Marsh, P. (1983) 'The West builds up its metals mountain', *New Scientist* (3 March 1983) pp. 573–7.
Mayo, W. (1979) 'Rent royalties', *The Economic Record*, vol. 55, no. 150, pp. 202–13.
McGill, S. C. (1981) 'Some issues and considerations for governments involved in formulating mineral agreements', Conference on Resource Potential and Implications of Mineral Development in the South Pacific Arovo Island, Papua New Guinea.
Mikesell, R. F. (1976) 'Financial considerations in netotiating mining development agreements', in Negotiating and draftings of mining development agreements', *Mining Journal Books*, London, pp. 17–20.
Mikesell, R. F. (1976) 'Rate of exploitation of exhaustible resources. The case of an export economy', *Natural Resources Forum*, vol. 1, no. 1, pp. 39–46.

Mikesell, R. F. (1980) 'Options for packaging agreements to next host government and foreign investment financial goals', paper presented at the *East-West Centre Workshop on Mineral Policies to Achieve Development Objectives*, Honolulu (June 9–13).
Mikesell, R. F. (1983) *Foreign Investment in Mining Projects* (Cambridge, Mass.: Oelyeschlager, Gunn and Hain).
Morrissett, I. (1967) 'Economic theory and resource policy', in Gaffney, M. (ed.) *Extractive Resources and Taxation* (University of Wisconsin Press) pp. 295–314.
Page, W. (1976) 'Mining and development. Are they compatible in South America?', *Resources Policy*, p. 235.
Palmer, K. F. (1980) 'Mineral Taxation policies in developing countries – an application of Resource Rent Tax', *International Monetary Fund Staff Papers*, vol. 27, no. 3, pp. 517–42.
Persaud, T. (1976) 'Conflicts between multinational corporations and less developed countries. The case of bauxite in the Caribbean with special reference to Guyana', Ph.D. thesis, Texas Tech. University.
Peterson, F. M. (1977) 'The government role in mineral exploration', in M. Crommelin and A. R. Thompson (eds) *Mineral Leasing as an Instrument of Public Policy* (Vancouver: University of British Columbia).
Parsons, R. B. and J. P. Gilbert (1979) 'Canadian mineral taxation', *Industrial Minerals*, October 1979.
Radetzki, M. (1977) 'Where should developing countries' minerals be processed? The country view versus the multinational company view', *World Development*, vol. 5, no. 4, pp. 325–34.
Radetzki, M. and S. Zorn (1979) 'Financing mining projects in developing countries', *Mining Journal Books*, London.
Reimer, H. (1977) 'Rates of return in the mining industry', *Can. Inst. Min. Metall.*, vol. 70, no. 785, pp. 101–5.
Roberts, W. (1967) 'Mine taxation in developing countries', in Gaffney, M. (ed.) *Extractive Resources and Taxation* (University of Wisconsin Press) pp. 197–218.
Schanze, E. (1980) 'Modes of government participation and control – an empirical assessment', *East-West Centre Workshop on Mineral Policies to Achieve Development Objectives*, Honolulu.
Schanze, E. et al. (1981) *Mining Ventures in Developing Countries, Analysis of Project Agreements*, Studies in Transational Law and Natural Resources (vol. 2) (Frankfurt am Main: Alfred Metzner Verlag).
Seers, D. (1963) 'Big companies and small countries. A practical proposal', *Kyklos*, vol. 16, pp. 599–608.
Sideri, S. and S. Johns (1980) *Mining for Development in the Third World: Multinational Corporations, State Enterprises and the International Economy* (Oxford: Pergamon Press, in co-operation with the Institute of Social Studies, The Hague).

Smith, D. N. (1980) 'New eyes for old: The future, present and past in the evolution of mineral agreements', *Harvard Law School, East-West Centre Workshop on Mineral Policies to Achieve Development Objectives*, Honolulu.

Smith, D. N. and L. T. Wells (1975) *Negotiationg Third World Mineral Agreements: Promises as Prologue* (Cambridge, Mass.: Ballinger).

Stevens, D. (1978) 'Foreign private investment – guidelines for assessing proposals in smaller developing countries', Commonwealth Secretariat, London.

Takeuchi, K., Thietach, C. and J. Hilmy (1977) 'Investment requirements in the non-fuel mineral sector in the developing countries', *Natural Resources Forum*, vol. 1, no. 3.

Thoburn, J. (1981) *Multinationals, Mining and Development – A Study of the Tin Industry* (Farnborough: Gower Press).

United Nations (1970) 'Mineral resources development with particular reference to the developing countries', UN Dep. of Economics and Social Affairs, ST/ECA/123 (New York).

United Nations (1972) 'Small scale mining in the developing countries', UN, ST/ECA/155, 171 (New York).

United Nationas (1974) 'The impact of multinational corporations on development and on international relations', Technical papers: Taxation (ST/ESA/11) (New York).

United Nations (1975) 'Measures to expand processing of primary commodities in developing countries', UNCTAD (TD/B/C17/197).

United Nations Centre for Transnational Corporations (1982) *Main Features and Trends in Petroleum and Mining Agreements.*

United Nations Centre for Transnational Corporations (1982) *Transnational Corporations and Contractual Relations in World Uranium Industry.*

United Nations Committee on Natural Resources (1982) *Permanent Sovereignty over Natural Resources.*

United Nations Industrial Development Organisation (UNIDO) (1978) *Mineral Processing in Developing Countries* (New York).

Walde, T. (1977) 'Lifting the veil from transnational mineral contracts. A review of recent literature', *Natural Resources Forum*, vol. 1, p. 277.

Walde, T. (1983) 'Permanent sovereignty over natural resources – recent developments in the mineral sector', *Natural Resources Forum*, vol. 7(3), pp. 239–51.

Walrond, G. W. (1976) 'Is scarcity of natural resources impeding economic growth?' *Can. Inst. Min. Metall.*, vol. 69, no. 772, pp. 114–19.

Whitney, J. W. and R. E. Whitney (1979) *Investment and Risk Analysis in the Minerals Industry* (Reno, Nevada).

Whitney, J. W. and R. E. Whitney (1980) 'A Comparative Analysis of the Mining tax burden of South Africa, the United States and Canada', *East-West Centre Workshop on Mineral Policies to Achieve Development Objectives*, Honolulu (June 9–13).

Zorn, S. A. (1977) 'New development in third world mining agreements', *Natural Resources Forum*, vol. 1, no. 3, pp. 239–50.

Zorn, S. A. (1981) 'Mineral processing in the Pacific island countries: problems and opportunities', Conference on Resource Potential and Implications of Mineral Development in the South Pacific Arovo Island, Papua New Guinea.

OFFICIAL PUBLICATIONS

Australia
Australian Taxation of Income from Foreign Investment and Natural Resources Industries.
Mining Regulations 1981 Government Gazette, Western Australia, Friday 13 November 1981.
'Mining giants come under fire', *Financial Times*, 4 September 1980.
Amendment to the Mining Act, Mineral Tr. Notes, 1978, vol. 75, no. 12, pp. 15–16.

Botswana
Mines and Minerals Act, 1967.

Canada
Natural Resources Department Act.
Mining Act and Mining Duties Act.
Government of Quebec, 1971.

Synopsis of Selected Income Tax Act.
Provisions directly relating to the Mineral Industry, September 1972.
Mineral Resources Branch, EMR.
Tibbo, L. (1981) *The Canadian Mineral Industry: Review and Outlook*, by Lic Tibbo, *Canadian Mineral Survey*, Energy, Mines and Resources.

Quebec's New Mining Laws, vol. 289. no. 7416, p. 289, 1977.

Malaysia
Mining Legislation, Malaysia.
Fourth Malaysia plan, 1981–85.
Economic Report, 1982–83, Ministry of Finance.

Papua New Guinea
Mining Act, 1977, No. 8 of 1978.
Financial Policies relating to Mining and Mining Tax Legislation, Statement of Intent, October 1977.

Sierra Leone
An Act to ratify and confirm an agreement for the prospecting and mining deposits of aluminium ores in certain areas of Sierra Leone, 1961.
Revised Mines Manual, 2nd Impression, Government of Sierra Leone.
The Legislative Framework – The Minerals Act (Cap. 196).
Income Tax Act, Ch. 273, 1969.
'An introduction to income tax in Sierra Leone', Income Tax Department of Sierra Leone (1973).

Tanzania
The Mining Act, 1979.
The Income Tax Act, 1973; The Income Tax (Amendment) Act, 1980.

Zambia
The Mines and Minerals Act, 1969, 1976.
The Prescribed Minerals and Materials Act, 1976.
Income Tax Act, 1968, 1976.

Index

Accelerated depreciation, *see* capital allowances
Accessional system, 9
Additional profits tax, *see* rent resource tax
Administration, of mining, 2, 9–33, 110; of tax collection, *see* fiscal instruments
Agreements, 87–110; concession style, *see* concession; double taxation, *see* double taxation; joint venture, *see* joint venture; mining, 1, 2, 55, 92, 104, 105; management, *see* management; negotiation of, *see* negotiation; production sharing, *see* production sharing; service, *see* service agreement; licensing, *see* licences; standardisation of, 6
Aluminium, 53, 86, 91, 96, 99
Angola, 53
Area, designated, 26; relinquishment of, 11, 23; rentals, 21; size of, 11, 21, 25, 27
Argentina, 53
Asante, S., 36
Asbestos, 6, 40(t)
Assignment of claims, 11, 21, 24

Barytes, 40(t)
Bauxite, 4, 5, 7, 40(t), 86, 88, 89, 90–1, 95, 96–7, 104, 124–35
Board, 93, 94, 119
Bonanza discovery, 37, 52–3
Bosson, R., 108
Bromwich, M., 176
Building products, 10–11, 20

Cadmium, 46

Calvo Principle, 105
Capital, 1, 24, 25, 26, 36, 53, 54, 56–86; foreign, 88, 122; capital intensity, 1, 36, 106–8; local, 111; movement of, 106; write-off, *see* capital allowances
Capital allowances, 34, 36, 44–5 (t), 47–50, 60–86, 96–103, 104, 107, 167–70; depreciation methods, 47–8, 164–66; (*see also* fiscal regime, financial regime)
Capital write-off, *see* capital allowances
Carried interest, 56–86, 94, 118
Claims, 24; size of, 11, 18; (*see also* assignment)
Coal, 7, 40(t), 56
Cobalt, 5, 46
Colombia, 53
Commonwealth, 2, 9; Commonwealth Secretariat, xv, 39, 105, 109–10
Company, 10–17, 89: incorporated, 11, 13, 29; joint management, 94; unincorporated, 11, 13, 29; vertically integrated, 107–9
Computing; computer-assisted financial capability, xiii, 3, 110; project analysis, 3, 56–86, 136–63
Concession, 10–17, 87–100, 107; joint venture, 89 (*see also* joint venture)
Conservation, 35–7, 42, 114; (*see also* depletion)
Contract, *see* agreement
Control, 93–5, 117, 119–21; of mining industries, 36, 72
Copper, 5, 6, 41(t), 46, 56, 90, 91, 94, 100–3, 104, 124–35
Corporate tax, 34, 43–52, 56–86 (*see also* income tax, fiscal regime, financial regime)

Corporation, *see* company
Costs, 53–4; cost–benefit calculations, 57; production, 39, 42, 43, 53–4, 56–86, 167–70; marginal cost, 42; minimum expenditure requirement, 11, 21; opportunity cost, 118
Custom duties, 34, 55, 96–103, 104, 113 (*see also* fiscal regime, financial regime)
Cut-off grade, 36, 42 (*see also* high grading)

Debt–equity ratio, 52, 54, 58
Depletion, depletionary, 5, 37 (*see also* conservation)
Deposits: development of ore, 20; geological definition of 20; size of deposits, 113, 116
Depreciation, *see* capital allowances
Development: economic development, 35, 37, 105; development demands, 105, 113; development tax, 45–6; mining development, 2, 107 *see also* mining; state of, 112
Diamonds, 4, 5, 41(t), 90–1, 96–103, 104; diamonds profit tax, 44–6, 52, 124–6
Disputes, Mining Affairs Appeal Tribunal, 31; resolution of, 15, 105, 112; settlement of, 15, 31, 107
Distributable profits, *see* profitability
Distribution: of consumption, 37; of profits, *see* profitability, economic rent; of project surplus, 39, 42, 56–86 (*see also* project)
Diversification, 6
Dividends, 57–86, 116; remittance, 36, 50 (*see also* profitability, withholding taxes)
Double taxation, 49, 89 (*see also* tax credibility)
Duties, *see* custom duties

Economic development, *see* development

Economic rent, 38–9, 42–3, 110, 114, 119 (*see also* rent resource tax, financial regime, fiscal regime)
Efficiency: of fiscal instruments, 52; operational, 38, 42–3
Employment, 4–7, 57 (*see also* Labour)
Enclave, 36 (*see also* regional development)
Endowment: human, 111; physical, 111, 112
Environmental protection, 33
Equity, 37, 51–2, 57–86, 93–5; control, 94; debt provision, 92; free equity, 56–63, 92–118 (*see also* return to equity, participation, ownership)
Exchange rate, 1, 167; effects, 57; exchange control, 104
Exhaustible, 37, 39, 46, 113; longetivity of supplies, 108 (*see also* conservation)
Expenditure, *see* costs
Exploration, 9–33, 46, 47–8, 50, 53, 96–103, 107
Exports: earnings, 4–7, 8(t); mining, 4–7, 8(t); re-exports, 7
Export tax, 34, 101

Feasibility study, 24
Fees, 34, 96–103; front-end, 36 (*see also* financial regime, fiscal regime)
Finance, financing 115; mining, 1, 43, 51–2, 77, 115, 118 (*see also* financial regime)
Financial model, *see* financial regime
Financial regime, 33–55, 56–86, 95–104, 115–19; definition, 34; model, 34, 56–86 (*see also* fiscal regime and fiscal instruments)
Fiscal instruments, 34, 35–55, 57–86, 95–104, 124–35; collection, 49–50, 53–4 (*see also* fiscal regime, financial regime)
Fiscal provision, *see* fiscal regime
Fiscal regime, 33–55, 57–86, 95–104, 115–119; equity 48; neutrality,

37, 60; stability, 43 (*see also* fiscal instruments)
Foreign exchange: availability, 87, 120; earnings, 4, 38, 46, 55 (*see also* export earnings); remittance, 36
Foreign investment, *see* investment
France, 108
Free equity, *see* equity
Freeman, Peter, 92
Front-end levies, *see* fiscal regime, fees, royalty

Gearing, *see* debt-equity ratio
Ghana, 53
Gillis, M., 54
Girvan, N., 36
Gold, 4, 6–7, 10, 15, 19, 38; model for gold project, *see* financial regime; price, 7; royalty, 41(t), 124–35
Govett, M. H., 94
Grants, 22–5
Gross domestic product, 4–7, 8(t)
Group of 77, 1
Guarantees: kind of, 23; performance, 11–15, 112
Guyana, 53, 88, 93

Herfindahl, O. C., 37
High grading, 42 (*see also* cut-off grade)
Holland, 49

Ilmenite, 98, 124
Import tax, *see* custom duties
Incentives, 34, 36, 38
Income, *see* revenue
Income tax, 34, 43–54, 57–83, 96–103, 113, 124–35, 167–70 (*see also* corporate tax)
Indonesia, 53
Industrial projects, *see* project
Inflation, 36, 37
Infrastructure, 38; geological, 111
Integration, 35–7, 120; economic linkages, 37, 38, 57, 105
Interest, 51–2, 54, 58–86, 96–103, 167 (*see also* withholding taxes)

Internal rate of return: definition, 176; calculations, 56–86, 115–17, 136–63 (*see also* returns)
Investment, 2, 35, 53; foreign, 2, 29, 82, 112; of mining, *see* mining; portfolio, 88 (*see also* investor)
Investor, 34, 36, 53, 88, 112; return to investor, *see* return to equity (*see also* investment)
Iron ore, 4, 5, 6, 7, 91; and steel, 4
Italy, 108

Japan, 108
Joint venture, 56–63, 88, 92, 119
Joint management company, *see* management

Kaolin, 6, 40(t)
Kneese, A. V., 37
Know-how, 1, 36, 54
Kumar, R., 47, 165

Labour: intensive, 18, 112; labour-surplus economy, 106; skills, 20
Land, 2; mining land, *see* mining; land tax/rent, 55
Land rent/tax, *see* land
Law of the sea, 1
Lead, 5, 41(t)
Lease, leasing, 11–15, 54; dredging, 27; general purpose, 26; mining, 20; sluicing, 27; underground, 27
Levies, *see* fiscal regime, fees
Levy, H., 166
Licences: 10–33; alluvial gold mining development, 20; duration of, 23; exclusive prospecting, 20; exploration, 20, 112; mining, 20, 26; period of, 20; prospecting, 20, 21; reconnaissance, 20; renewal of, 22, 25; succession in, 22
Linkages, *see* integration
Localisation: guidelines, 29; requirements, 121; local participation, 10–17, 29–30, 89
Location, 36, 107

Management, 9–33, 93–5, 119–21; joint management company, 94; management agreement, 92
Manganese, 40(t)
Marginal cost, *see* costs
Marginal project, *see* project
Marginal returns, *see* returns
Marketing, 10, 104–5
Marsh, P., 108
Mica, 40(t)
Mineral: alluvial, 19, 20; area, 24; building, 18, 27, 29; definition of, 19; deposits, 118; development in less developed countries, 87, 88, 93, 112; exclusivity over, 24, 25; exhaustible non-renewable nature of, 113; exhaustible stock, 120, 122; high risk development of, 119; industrial, 18, 19, 25, 27, 29; large-scale activity, 29, 105; radioactive, 19 (*see also* mining *and* reference to each kind of mineral)
Mineral sands, 7
Mining; agreements, *see* agreements; alluvial, 26; artisinal, 106, 113; building products, 26; capital, *see* capital; characteristics of, 106; code, 6 (*see also* statutes); development, 2; employment, 4–7; establishments (companies), 7; fiscal regime, 33–35; finance, *see* finance; land, 2, 52; lease, *see* lease; licence, *see* licences; legislation, 2, 11, 20, 26, 38; legislative provisions, 9–33; policy, 1, 3, 6; rights, 11, 26; scale of, 11, 18; sector, 4, small-scale, 26, 29, 38, 113; special casement, 116, 117; tax, *see* taxation; tribunal, 31
Multinational, 3, 42, 54, 86, 107–9 (*see also* transfer, pricing)

Nationalisation, 89, 94
Negotiation: of mining agreements, 54, 86, 87, 109–10, 121
New International Economic Order, 1
Nickel, 5, 7, 90–1
Non-aligned Movement, 1

Non-renewable, *see* exhaustible
Non-replenishable, *see* exhaustible
North–South Dialogue, 1
Norway, 54

Oil: industry, 5; price, 1; production bonuses, 34; tax, 53, 54; inclusion of hydrocarbons in legislation, 10–17
Opportunity cost, *see* cost
Output, *see* production
Ownership, 46; government, 2; majority-local, 95; nominal, 119; rights to, 112 (*see also* equity, participation)

Parliamentary ratification, 15, 113
Participation; government, 15, 34, 37, 57–86, 89, 90–1, 92, 96–103, 117; joint, *see* joint venture; local, 25, 29, 30, 89 (*see also* equity, ownership)
Payback period, 49, 53
Payroll tax, 55
Per capita income, 4–7
Performance guarantee, *see* guarantees
Permits, 10–17; mineral, 27; mining, 18; reconnaissance, 20; restricted minerals, 27
Petroleum, *see* oil
Phosphate, 5
Platinum, 41(t)
Policy of government, 1, 3, 6, 35–9
Pollution, 57
Prices, 42, 56–7; of minerals, 43; price expectations, 36–7; pricing in agreements, 104–5 (*see also* transfer pricing)
Price expectation, *see* prices
Production, 56–86; of minerals, 39, 42; value, 43
Production bonus, 34
Production sharing, 57, 92, 93
Profitability, 36, 50; as related to taxation, 39–42, 52–4, 56–86, 113; distributable profits, 92 (*see also* dividends)

Index

Project: analysis, 3, 56–86, 136–63; development, 115; economics, 34; industrial, 20; marginal, 36, 37, 52, 78–86, 116; rate of return, *see* returns; surplus, 36, 115 (*see also* distribution)
Property tax, 53

Quarry, quarrying, 5

Radetzki, M., 109
Rate of return, *see* returns
Regional development, 55, 57 (*see also* enclave)
Reinvestment, 50, 84
Relinquishment, 55
Rentals, 21, 124–35
Rent resource tax, 34, 46, 49, 52–4, 61–86, 94, 129, 167–70 (*see also* economic rent)
Repatriation of funds, 105
Returns: marginal returns, 115; project rate of return, 56–85, 115, 116, 117 (*see also* internal rate of returns)
Return to equity, 34, 49, 51, 53–4, 56–86
Revenue, 4, 7, 8(t), 35–55, 56–86, 136–63, 167–70; revenue maximisation, 35–6, 42
Rights: 10–17, accessional, 10; of aliens, 29; ancestral, 9, 10; customary, 27; domanial, 10(t); exclusive, 22, 24, 25; mineral, 9, 24, 28, 31, 88, 89; minority, 94; non-exclusive, 22; preferential, 25; property, 30, 107, 111, 113; prospecting, 11, 20, 26, 30; relinquishment of, 22, 23; surface, 9, 26, 28; termination of, 30, 122; to work, 24
Risk, 1, 46, 47, 52, 54, 77, 78, 86, 119, 121
Robinson, H. A., 94
Royalty, 34, 36, 39–43, 50, 52, 53, 54, 57–86, 96–103, 113, 114, 124–35, 167–70; front-end, 42–3
Rutile, 4, 40(t), 89, 91, 94, 96–9, 124

Sales tax, 46, 52, 55
Sarnat, M., 176
Scale, *see* mining
Selenium, 46
Service agreement, 92
Shipping, 6
Signature bonus, 34
Silver, 41(t), 46
Size, *see* area, claims, deposits
Small-scale mining, *see* mining
Smuggling, 38
South Korea, 108
Sovereignty, 2, 34, 37, 38, 46, 72, 86, 87, 105, 110, 111, 115, 119, 121, 122
Spain, 108
Stamp duties, 55
Statutes: mining, 2, 3, 9–33, 35, 37, 57, 107 (*see also* agreements, code)
Stockpiling, 108
Surplus, *see* profitability, project
Surtax, 34, 44–5(t), 46
Sweden, 108

Tantalite-columbite, 7
Tax collection, *see* fiscal instruments
Tax credibility, 54, 108–9, 114 (*see also* double taxation)
Tax holiday, 34, 36, 48–50, 53, 85, 104, 107
Tax-sparing provisions, 49
Taxes, *see under individual type of taxes, and see also* fiscal instruments, fiscal regime, financial regime
Technology: monopoly of, 1
Thoburn, J., 94
Thorium, 41(t)
Tin, 4, 5, 6, 9, 40(t), 56, 94, 130
Titanium, 91
Training, 32–3, 38, 105, 110
Transfer pricing, 42, 54, 104–5, 109, 114, 120, 121, 167 (*see also* price)
Tungsten, 41(t)
Turn-key, 57

United Nations Department for Technical Cooperation and Development, 109

United Nations Development Programme (UNDP), 111
United Nations Industrial Development Organisation (UNIDO), 109
United Kingdom, 53, 54, 108
Uranium, 4, 90, 99

Value added, 5, 7, 8(t), 120
Varon, R., 108

Walrond, G., 36, 47, 165
Wells, L. T., 54
Windfall, *see* bonanza
Withholding taxes, 44–5(t), 50–2, 57–86, 96–103, 116, 124–35
Works: 10–17; value of, 23; minimum obligation, 11, 23; surface for ancillary works, 26

Zinc, 5, 41(t)
Zircon, 41(t)

GPSR Compliance
The European Union's (EU) General Product Safety Regulation (GPSR) is a set of rules that requires consumer products to be safe and our obligations to ensure this.

If you have any concerns about our products, you can contact us on

ProductSafety@springernature.com

In case Publisher is established outside the EU, the EU authorized representative is:

Springer Nature Customer Service Center GmbH
Europaplatz 3
69115 Heidelberg, Germany

www.ingramcontent.com/pod-product-compliance
Lightning Source LLC
Chambersburg PA
CBHW031541230426
43749CB00025B/437